D0927833

n-gons

Friedrich Bachmann and Eckart Schmidt
Translated by Cyril W.L. Garner

This book, a translation of the German volume *n-Ecke*, presents an elegant geometric theory which, starting from quite elementary geometrical observations, exhibits
an interesting connection between geometry and fundamental ideas of modern algebra
in a form that is easily accessible to the student who lacks a sophisticated background
in mathematics. It stimulates geometrical thought by applying the tools of linear
algebra and the algebra of polynomials to a concrete geometrical situation to reveal
some rather surprising insights into the geometry of *n*-gons. The twelve chapters treat
n-gons, classes of *n*-gons, and mappings of the set of *n*-gons into itself. Exercises are
included throughout, and two appendixes, by Henner Kinder and Eckart Schmidt,
provide background material on lattices and cyclotomic polynomials. (Mathematical
Expositions No. 18)

FRIEDRICH BACHMANN is at Christian Albrechts University in Kiel.
ECKART SCHMIDT is also in Kiel.
CYRIL W.L. GARNER is associate professor in the Department of Mathematics at Carleton
University, Ottawa.

MATHEMATICAL EXPOSITIONS

Editorial Board

H.S.M. COXETER, G.F.D. DUFF, D.A.S. FRASER,
G. de B. ROBINSON (Secretary), P.G. ROONEY

Volumes Published

1 *The Foundations of Geometry*   G. de B. ROBINSON
2 *Non-Euclidean Geometry*   H.S.M. COXETER
3 *The Theory of Potential and Spherical Harmonics*
   W.J. STERNBERG and T.L. SMITH
4 *The Variational Principles of Mechanics*   CORNELIUS LANCZOS
5 *Tensor Calculus*   J.L. SYNGE and A.E. SCHILD
6 *The Theory of Functions of a Real Variable*   R.L. JEFFERY
7 *General Topology*   WACLAW SIERPINSKI
   (translated by C. CECILIA KRIEGER) (out of print)
8 *Bernstein Polynomials*   G.G. LORENTZ (out of print)
9 *Partial Differential Equations*   G.F.D. DUFF
10 *Variational Methods for Eigenvalue Problems*   S.H. GOULD
11 *Differential Geometry*   ERWIN KREYSZIG (out of print)
12 *Representation Theory of the Symmetric Group*   G. de B. ROBINSON
13 *Geometry of Complex Numbers*   HANS SCHWERDTFEGER
14 *Rings and Radicals*   N.J. DIVINSKY
15 *Connectivity in Graphs*   W.T. TUTTE
16 *Introduction to Differential Geometry and Riemannian Geometry*
   ERWIN KREYSZIG
17 *Mathematical Theory of Dislocations and Fracture*   R.W. LARDNER
18 *n-gons*   FRIEDRICH BACHMANN and ECKART SCHMIDT
   (translated by CYRIL W.L. GARNER)
19 *Weighing Evidence in Language and Literature: A Statistical Approach*
   BARRON BRAINERD

MATHEMATICAL EXPOSITIONS NO. 18

n-gons

FRIEDRICH BACHMANN

Christian Albrechts University, Kiel

ECKART SCHMIDT

Kiel

Translated by

CYRIL W.L. GARNER

Carleton University, Ottawa

UNIVERSITY OF TORONTO PRESS

WITHDRAWN

LIBRARY OF ST. MARY'S COLLEGE EMMITSBURG, MARYLAND

© University of Toronto Press 1975
Toronto and Buffalo

Printed in Great Britain

ISBN 0–8020–1843–2
CN ISSN 0076-5333
LC 70–185699
AMS 1970 subject classifications
50–D05, 06–A40, 15–A03, 12–E10

To H.S.M. Coxeter

Contents

Author's preface   xi

Translator's preface   xiii

Summary of contents   xv

Introduction   3

Part I
Cyclic classes and cyclic mappings   11

1 Cyclic classes of n-gons   13
    1 n-gons, vector space of n-gons   13
    2 Cyclic classes   14
    3 The centre-of-gravity n-gon. The zero isobaric class   16
    4 Two types of cyclic classes   17
    5 Periodic classes   19
    6 Degree of freedom of a cyclic class   20
    7 Dimension of an n-gon   21
    8 Examples of cyclic classes   22

2 Cyclic mappings of n-gons   32
    1 Cyclic mappings   32
    2 The algebra of cyclic mappings   33
    3 Coefficient sum of a cyclic mapping   35
    4 Projections   36
    5 Examples   40
    6 A cyclic quasi-projection   45
    7 Isobaric cyclic projections for $n = 4$ 47
    8 Cyclic matrices   49

3 Isobaric cyclic mappings   51
    1 σ-kernel   51
    2 Two types of cyclic classes   52

3 Remarks concerning isobaric cyclic mappings 55
Notes concerning the addition of n-gons 56
1 Addition of n-gons 56
2 Displacement of addition in an abelian group 57
3 Addition of isobaric n-gons with respect to the centre of gravity 58

4 Averaging mappings 60
1 Isobarically splitting n-gons 60
2 Omitting averaging projections 61
3 The complementary projections 62
4 Consecutive averaging projections 64

PART II
The main theorem 67

5 Idempotent elements and Boolean algebras 69
1 Idempotent elements of a ring 69
2 Finitely generated Boolean algebras 71
3 Idempotent endomorphisms of an abelian group: Im-transfer 74
4 The Boolean algebra of cyclic projections 76
5 Examples of Im-transfer 77

6 The main theorem about cyclic classes 81
1 Congruences in principal ideal domains 81
2 Main theorems about cyclic mappings and cyclic classes 84
3 The prime factors of $x^n - 1$ and the atomic cyclic classes 89

PART III
Boolean algebras of n-gonal theory 93

7 Idempotent-transfer. Residue class rings of principal ideal domains 95
1 R-modules 95
2 Idempotent-transfer 96
3 A special case of idempotent-transfer 96
4 Ideals and divisibility in a principal ideal domain 97
5 Residue class rings of principal ideal domains 98
6 Residue class rings as sums of residue class rings 102

8 Boolean algebras of the n-gonal theory I 104
1 The Boolean algebras L_1-L_5 104
2 Divisors of $x^n - 1$ and cyclic classes 108
3 Spectrum 110
4 Examples of defining cyclic classes by divisors of $x^n - 1$ 112

9 Boolean algebras of the *n*-gonal theory II 116
   1 Galois correspondence of annihilators and kernels 116
   2 Ideal-transfer 118
   3 Second proof of the main theorem: Main diagram 119
   4 Graduation: Degree of freedom of a cyclic class 122
   5 Miscellaneous exercises 125

PART IV
Atomic decompositions 129

10 Rational components of an *n*-gon 131
   1 Q-regular *n*-gons 131
   2 Cyclic classes defined by cyclotomic polynomials 134
   3 Rational components of an *n*-gon 136
   4 The Boolean algebra generated by omitting averaging
      projections and its atoms 138
   5 Construction of the rational components of an *n*-gon 139

11 Complex components of an *n*-gon 141
   1 *w*-*n*-gons, regular *n*-gons 141
   2 The case of the complex field 143
   3 Complex components of an *n*-gon 145

12 The real components of an *n*-gon 149
   1 Anticyclic cyclic classes 149
   2 A special type of cyclic systems of equations 151
   3 Affinely regular *n*-gons 154
   4 Three extreme cases for the Boolean algebra of cyclic *n*-gonal classes 158
   5 Real components of an *n*-gon 160

Appendices 165

   1 Lattices HENNER KINDER 167

   2 Cyclotomic polynomials ECKART SCHMIDT 179

List of symbols and notations 187

Index 189

Authors' preface

Mathematics possesses an ability to extract new facets from things which surround us and to extend our intuitions in unexpected ways.

P. ALEXANDROFF

This book is about n-gons, classes of n-gons, and mappings of the set of n-gons into itself.

n-gons are particularly primitive geometric objects; everyone knows special n-gons. The geometric observations concerning, for example, hexagons and classes and mappings of hexagons, from which the questions of this book originate, are so elementary that they can be discussed even with the mathematically uninitiated.[1] General questions are answered by algebraic means.

In order to provide an easy approach to the algebraic treatment, we define an n-gon as an n-tuple

$$(a_1, a_2, ..., a_n)$$

of elements of a vector space over a field (in which $n \cdot 1 \neq 0$). With the concept of the vector n-tuple we find ourselves from the very beginning on what G. Choquet calls the 'Royal Road' of linear algebra.[2]

Certain sets of n-gons, called cyclic classes, are one principal object of the investigation. A prototype of a cyclic class is, for $n = 4$, the class of parallelograms: the set of quadrangles (a_1, a_2, a_3, a_4) for which $a_1 - a_2 + a_3 - a_4 = o$. In general, a cyclic class of n-gons consists of all n-gons which satisfy a 'cyclic' system of homogeneous linear equations with coefficients from the given field.

The main theorem concerning cyclic classes says that the number of cyclic classes is finite for each n, or, more precisely, that the cyclic classes form a finite Boolean algebra. Every n-gon is uniquely decomposable into n-gons from 'atomic' cyclic classes. Accordingly n-gons possess an atomic structure. The building blocks – the n-gons of the atomic cyclic classes – are distinguished by regularity.

The algebraic tools which we use lie in the mainstream of algebra, and one concern of this book is to exhibit what we think to be an exciting connection between the geometry of n-gons and ordinary algebra.

---

1 See the introduction and the diagram of the eight hexagonal classes in §§1.8 and 2.5.
2 G. Choquet, *Geometry in a Modern Setting* (London, 1969); trans. from the French text *L'Enseignement de la géométrie* (Paris, 1964).

We have assumed that the reader is familiar with the basic concepts of linear algebra and of algebra, especially with concepts such as groups, rings, fields, vector spaces, and homomorphisms.[3] Concepts and simple facts from the theory of lattices and properties of cyclotomic polynomials are collected in two appendices.

We wish to express our thanks particularly to H. Kinder, who is really one of the authors of the theory developed here, for many discussions and also for the appendix on lattices which he contributed to this book; and to U. Spengler, who checked the 126 exercises. We are also indebted to both of them, and to L. Broecker and P. Klopsch, for their suggestions and help with the proofreading.

During the writing of these lectures we have often thought with gratitude of the encouragement given by all those mathematicians who have found pleasure in this little theory, either in lectures or in private talks, and who have urged us to publish it.

F.B.
E.S.
Kiel, September 1969

---

3 For example, the following texts are recommended: S. MacLane and G. Birkhoff, *Algebra* (New York, 1967); H.J. Kowalsky, *Linear Algebra* (Berlin, 1965); S. Lang, *Linear Algebra* (Reading, Mass., 1966); R. Kochendoerffer, *Einführung in die Algebra* (Berlin, 1955); B.L. van der Waerden, *Modern Algebra* I (New York, 1949); trans. from the 2nd rev. German edition *Moderne Algebra* I (Berlin, etc., 1940).

Translator's preface

In this book the authors present an elegant geometric theory which, starting from quite elementary geometrical observations, exhibits an interesting connection between geometry and fundamental ideas of modern algebra. The reader is given an opportunity to apply his knowledge of linear algebra and the algebra of polynomials to a concrete geometrical situation, and thus obtain some rather surprising insights into the geometry of n-gons.

Such books are quite rare, and it seemed desirable to me that this book, written in German and recently translated into Russian, should also be accessible in English. The material has been used by Professor Bachmann for a university course in both Germany and the United States, and Mr Schmidt has adapted the more elementary parts for high school purposes.

C.W.L.G.
Ottawa, July 1973

Summary of contents

After the introduction the concept of a cyclic class of n-gons and the analgous concept of cyclic mappings are defined in chapters 1 and 2. Chapters 1–4 (part I) are concerned mainly with giving examples.[1]

In part II, chapter 5 and §6.1,[2] algebraic tools are developed with the help of which the main theorem concerning cyclic classes can be proved in §6.2.

In part III, after further algebraic preparations in chapter 7 and §§9.1–2, chapters 8 and 9 exhibit systematic approaches to the Boolean algebra of cyclic n-gonal classes. Figure 61 (p. 121) gives a summary.

Part IV, chapters 10–12, is concerned with the decomposition of an n-gon into the n-gons of the atomic cyclic classes when the underlying field is the field of rational, complex, or real numbers.

A possibility for a shorter course consists in reading only chapter 1, §§1–5 of chapter 2, §§1–4 of chapter 5, and chapter 6.

---

1 Here one could proceed more naïvely and, continuing in the style of the introduction, consider special classes and mappings of n-gons, without worrying about general concepts.
2 §6.1 means section 1 of chapter 6, and §1 refers to section 1 of the chapter being considered.

n-gons

ℵ-gons

Introduction

For introductory purposes let us assume a Euclidean space[1] and use intuitive language uncritically.

The following well-known theorem forms the starting-point for our theory of n-gons: *the mid-points of the sides of any quadrangle form a parallelogram* (figure 1).

Let us interpret this theorem as follows. To any quadrangle A there is associated a parallelogram $A°$ formed by the mid-points of the sides. This is a mapping of the set of quadrangles into itself. Moreover there is a special class of quadrangles, the class of parallelograms. The theorem says: the mapping maps the set of all quadrangles into the special class. Thus it 'specializes' the set of quadrangles. This specialization is also seen in a diminishing of the maximum dimension: the given quadrangle need not be planar, but a parallelogram is at most two-dimensional.

This initial theorem

(i) *If A is a quadrangle, then $A°$ is a parallelogram*

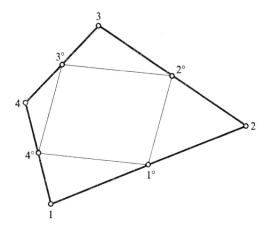

Figure 1

1 However we shall use only affine concepts and results, not metric.

immediately suggests the question whether the formation of the n-gon $A°$, whose vertices are the mid-points of the sides of the given n-gon A, is a specializing mapping for every n. We note that for $n = 3$ it does not specialize, and in general it does not specialize for odd n. On the other hand, a specialization[2] appears when $n = 6$, but it is not so striking as for $n = 4$. For even $n = 2m$ we have, in general,

(ii) *If A is a 2m-gon, then $A°$ is a 2m-gon whose alternate sides form a closed vector polygon.*

More precisely: if the sides of A are marked alternately by a and b, the a-sides form a closed vector polygon,[3] as do the b-sides. For example, in the Thomsen

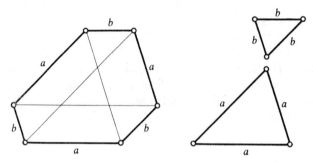

Figure 2

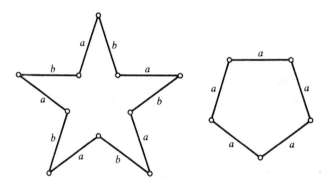

Figure 3

2 I.M. Yaglom, *Geometric Transformations* (Russian ed.: Moscow, 1955), p. 31; English trans. (New York, 1962).
3 If the a-sides are moved (in any order) by parallel displacements so that the end-point of one side coincides with the starting-point of the next, then the end-point of the last one coincides with the starting-point of the first.

hexagon[4] (figure 2) and the 5-pointed star of figure 3 alternate sides form closed vector polygons.

It is worth noting that the symbol 4 does not appear in the word 'parallelogram.' This suggests defining a class of $2m$-parallelograms in the set of $2m$-gons: a $2m$-gon is called a $2m$-parallelogram if every side forms, with its opposite side, a parallelogram. Figure 4 shows a 6-parallelogram. The outline of a matchbox projected onto a plane is a 6-parallelogram, but a 6-parallelogram can also be a three-dimensional figure.

If A is a hexagon, then $A°$ is not in general a 6-parallelogram. But we can look for mappings of the set of hexagons into itself which associate a 6-parallelogram with any hexagon.

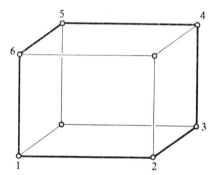

Figure 4

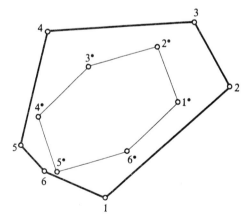

Figure 5

4 G. Thomsen, 'Schnittpunktsätze in ebenen Geweben,' *Abh. Math. Sem. Univ. Hamburg* 7, 99–106 (1930), fig. 3; and W. Blaschke and G. Bol, *Geometrie der Gewebe* (Berlin, 1938), fig. 28.

In the construction of the n-gon formed by the mid-points of the sides we took, for any two consecutive vertices of the given n-gon A, the mid-point, that is, the centre of gravity. We can carry this one step further and construct a new n-gon $A^{\bullet}$ from A by forming the centre of gravity of three consecutive vertices of A. (If the vertices of A are numbered 1, 2, ..., n, then we form the centre of gravity $1^{\bullet}$ of the vertices 1, 2, 3, the centre of gravity $2^{\bullet}$ of 2, 3, 4, etc., and let $A^{\bullet}$ be the n-gon with vertices $1^{\bullet}$, $2^{\bullet}$, ..., $n^{\bullet}$.) Then (figure 5),

(iii) *If A is a hexagon, $A^{\bullet}$ is a 6-parallelogram.*

Next, for any three consecutive vertices of a given n-gon A, form the fourth point of the parallelogram determined by them, called the fourth parallelogram point. (Thus form the fourth parallelogram point $1'$ of 1, 2, 3; the fourth parallelogram point $2'$ of 2, 3, 4; etc.) Then we obtain an n-gon A'. Figure 6 shows a triangle A with its triangle A'. Then (figure 7)

(iv) *If A is a hexagon, A' is a prism.*

Figures 8 and 9 illustrate what is meant by a prism. Note the numbering of the vertices. Theorem (iv) is particularly noteworthy since the mapping A to A' does not specialize for $n = 1, 2, 3, 4, 5, 7, 8, 9, 10, 11$.

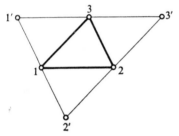

Figure 6

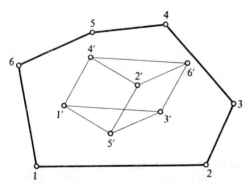

Figure 7

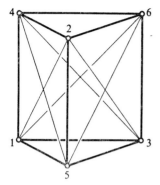

Figure 8

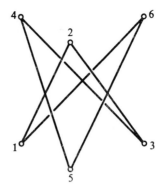

Figure 9

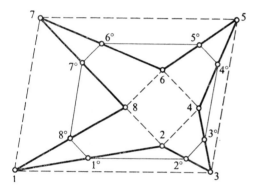

Figure 10

Let us mention one more theorem, this time with the given n-gon already belonging to a special class:

(v) *If A is an octagon whose two sub-quadrangles formed by alternate vertices are parallelograms, then $A°$ is an 8-parallelogram* (figure 10).

An octagon satisfying the conditions is obtained by choosing two arbitrary parallelograms (independent of each other), numbering their vertices 1, 3, 5, 7 and 2, 4, 6, 8 respectively, and then joining 1 and 2, 2 and 3, ..., 8 and 1.

By (ii), (iii), (iv) the set of hexagons is specialized by each of the mappings under consideration. This specialization can be carried further by performing two or three of the mappings successively. For example:

(vi) *If A is a hexagon, $A°•$ is an affinely regular[5] hexagon* (figure 11) *and $A°•'$ a 'trivial' hexagon, i.e. a point counted 6 times.*

5 See §1.8 for a definition of an affinely regular hexagon.

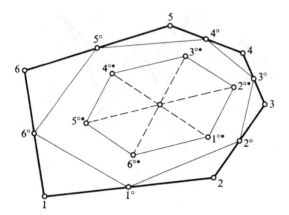

Figure 11

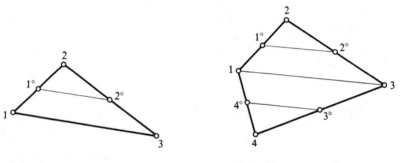

Figure 12   Figure 13

Here the maximum dimension is drastically reduced; a hexagon can be 5-dimensional, an affinely regular hexagon is at most 2-dimensional, and a trivial hexagon is 0-dimensional.

Now, how can we prove such theorems?

First, let us see how far we can go with a knowledge of elementary geometry. (i) follows from the theorem: the line segment joining the mid-points of two sides of a triangle is parallel to the third side, and its length is one-half of that side (figures 12, 13).

With this theorem, we can also prove (v). We can continue: in a quadrangle with vertices 1, 2, 3, 4 let $1^{\bullet}$, $2^{\bullet}$ be the centres of gravity of 1, 2, 3 and 2, 3, 4 respectively. Then the line segment determined by $1^{\bullet}$, $2^{\bullet}$ is parallel to the side with vertices 1, 4, and its length is one-third the length of that side (figure 14). Six applications of this theorem yield a proof of (iii).

Nevertheless we wish to use vector calculus. By a choice of origin we represent

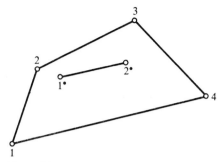

Figure 14

the points of space by position vectors o, a, b, ..., and an n-gon, as an n-tuple of its vertices, by an n-tuple of vectors

$(a_1, a_2, ..., a_n)$.

Then we merely have to express the concepts appearing in (i)–(vi) in the language of vectors, and the proofs become trivial. We no longer distinguish between a point and its representing vector, or between an n-gon and its representing n-tuple of vectors.

(a_1, a_2, a_3, a_4) is then a parallelogram if and only if $a_2 - a_1 = a_3 - a_4$; thus
$$a_1 - a_2 + a_3 - a_4 = o, \qquad (*)$$
i.e. the alternating sum of the vertices (vertex vectors) is o.

PROOF OF (i) Let $A = (a_1, a_2, a_3, a_4)$ be an arbitrary quadrangle. The quadrangle formed by the mid-points of the sides is

$A^\circ = (\tfrac{1}{2}(a_1 + a_2), \tfrac{1}{2}(a_2 + a_3), \tfrac{1}{2}(a_3 + a_4), \tfrac{1}{2}(a_4 + a_1))$.

The alternating sum of the vertices of this quadrangle is o; hence A° is a parallelogram.

In a $2m$-gon $(a_1, a_2 ..., a_{2m})$ the alternate sides form closed vector polygons when $(a_2 - a_1) + (a_4 - a_3) + ... + (a_{2m} - a_{2m-1}) = o$ and $(a_3 - a_2) + (a_5 - a_4) + ... + (a_1 - a_{2m}) = o$. Both of these equations can be written in the form

$$a_1 - a_2 + a_3 - a_4 + ... + a_{2m-1} - a_{2m} = o,$$

which says that the alternating sum of the vertices is o.

PROOF OF (ii) Let $A = (a_1, a_2, ..., a_{2m})$ be an arbitrary $2m$-gon. The $2m$-gon formed by the mid-points of the sides is

$A^\circ = (\tfrac{1}{2}(a_1 + a_2), \tfrac{1}{2}(a_2 + a_3), ..., \tfrac{1}{2}(a_{2m-1} + a_{2m}), \tfrac{1}{2}(a_{2m}, a_1))$.

The alternating sum of the vertices of this $2m$-gon is o.

That a hexagon $(a_1, a_2, ..., a_6)$ is a prism means that $a_1 - a_4 = a_3 - a_6 = a_5 - a_2$ (figure 8).

PROOF OF (iv)   Let $A = (a_1, a_2, ..., a_6)$ be an arbitrary hexagon. We denote the fourth parallelogram point of a_1, a_2, a_3 by a_1', the fourth parallelogram point of a_2, a_3, a_4 by a_2', etc. By theorem (*)

$$a_1 - a_2 + a_3 = a_1', \quad a_2 - a_3 + a_4 = a_2', \quad ..., \quad a_6 - a_1 + a_2 = a_6'.$$

For the hexagon $A' = (a_1', a_2', ..., a_6')$ we then have

$$a_1' - a_4' = a_3' - a_6' = a_5' - a_2',$$

for each of these differences is equal to $a_1 - a_2 + a_3 - a_4 + a_5 - a_6$. Thus A' satisfies the prism condition.

The verification of (iii), (v), (vi) is left to the reader.

Theorems (i)–(vi) are concerned with mappings which have obvious geometrical meanings, and state that for a definite n certain classes of n-gons are mapped into special classes. Certainly there are many other theorems of this type. In order to have a conceptual framework for general questions, we must first of all state definitely which sets of n-gons should be admitted as 'classes.' Likewise, we must decide which mappings of the set of n-gons into itself should be considered. The simple results of this introduction lead one to hope that we shall find deeper properties of n-gons by studying, under appropriate definitions, the interplay of classes and mappings.

From these examples we shall abstract the concept of a cyclic class of n-gons, and the concept of cyclic mappings. The definitions are given in chapters 1 and 2.

REFERENCES

F. Bachman, 'Punkte, Vektoren, Spiegelungen,' *Der Mathematische und Natur-wissenschaftliche Unterricht* (MNU), 18 (1965), 97–111.

F. Bachmann and J. Boczeck, 'Punkte, Vektoren, Spiegelungen,' *Grundzüge der Mathematik*, ed. H. Behnke, F. Bachmann, and K. Fladt (Göttingen, 1967), vol. II A, chap. 2, pp. 30–65.

Eckart Schmidt, 'Abbildungen und Klassen von n-Ecken,' MNU, 25 (1972), 146–56.

PART I
Cyclic classes and cyclic mappings

1
Cyclic classes of n-gons

1
n-GONS, VECTOR SPACE OF n-GONS

Let n be a natural number and K a (commutative) field whose characteristic does not divide the number n. The elements of K are denoted by $a, b, \dots$, the zero element by 0, and the unit element by 1. By the condition imposed upon the characteristic, the element $1/n$ exists in K. (This condition is fulfilled for every n in fields of characteristic 0, in particular the field Q of rational numbers.)

Let V be a vector space over K. The elements $a, b, \dots$ of V are also called *points*, and in particular the zero vector o is called the *zero point*. Let the dimension of V be non-zero (or else V would consist of the zero vector only) but otherwise arbitrary, finite or infinite. We do not assume that there is any connection between the number n and the dimension of V.

We call the n-tuples $(a_1, a_2, \dots, a_n)$ of elements of V n-gons. We also denote n-gons by $A, B, \dots$, and in particular the *zero n-gon* $(o, o, \dots, o)$ by O. $\mathfrak{A}_n$ denotes the set of all n-gons.

We define *addition* and *multiplication* by elements of K for the n-gons as follows:

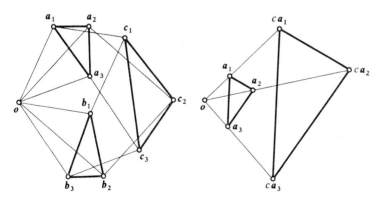

Figure 15

$$(a_1, a_2, ..., a_n) + (b_1, b_2, ..., b_n) = (a_1 + b_1, a_2 + b_2, ..., a_n + b_n),$$

$$c(a_1, a_2, ..., a_n) = (ca_1, ca_2, ..., ca_n).$$

(See figure 15.)

The set $\mathfrak{A}_n$ of n-gons is then a vector space over K, namely the vector space $V^n = V \oplus V \oplus ... \oplus V$, called the *vector space of n-gons*.

By definition an n-gon $(a_1, a_2, ..., a_n)$ is an ordered set of n points of V. Using geometrical language, the points $a_1, a_2, ..., a_n$ are called *vertices*, the ordered pairs (a_i, a_{i+1}) of consecutive vertices are called *sides* of the n-gon, and the differences $a_{i+1} - a_i$ *side vectors*. The indices are to be read modulo n so that (a_n, a_1) is also a side. If $n = 2m$, we call a_i and a_{m+i} *opposite vertices*, and (a_i, a_{i+1}) and (a_{m+i}, a_{m+i+1}) *opposite sides* of the $2m$-gon.

Moreover an n-gon $(a_1, a_2, ..., a_n)$ may be interpreted as an element of V^n, the vector space of n-gons.

In the definition of an n-gon $(a_1, a_2, ..., a_n)$, nothing was assumed about the distinctness of the vertices $a_1, a_2, ..., a_n$. We shall call an n-gon $(a, a, ..., a)$ a *trivial n-gon*. Since we can also consider a trivial n-gon in V as a 1-gon counted n times, we denote the set of trivial n-gons by $\mathfrak{A}_{1,n}$. Then $\mathfrak{A}_{1,n}$ is a subspace of the vector space of n-gons.

2

CYCLIC CLASSES

Given an n-gon $(a_1, a_2, ..., a_n)$ we can form the n-gons

$$(a_2, ..., a_n, a_1), \quad (a_3, ..., a_1, a_2), \quad ..., \quad (a_n, a_1, ..., a_{n-1}) \tag{1}$$

by cyclically permuting the vertices of the given n-gon.

Corresponding to the geometric meaning of an n-gon, we shall study sets of n-gons which contain all n-gons (1) if they contain an n-gon $(a_1, a_2, ..., a_n)$. In general, sets of n-gons $(a_1, a_2, ..., a_n)$ which satisfy a linear equation

$$c_0 a_1 + c_1 a_2 + ... + c_{n-1} a_n = o$$

with given coefficients $c_0, c_1, ..., c_{n-1} \in K$ do not have this property.

Consider now *cyclic systems of equations*, i.e. linear systems of homogeneous equations:

$$c_0 a_1 + c_1 a_2 + ... + c_{n-1} a_n = o,$$
$$c_0 a_2 + c_1 a_3 + ... + c_{n-1} a_1 = o, \tag{2}$$
$$\cdots\cdots\cdots$$
$$c_0 a_n + c_1 a_1 + ... + c_{n-1} a_{n-1} = o$$

with given coefficients $c_0, c_1, ..., c_{n-1} \in K$. The set of all n-gons which satisfy such a system of equations contains the n-gons (1) if it contains an n-gon $(a_1, a_2, ..., a_n)$.

By a cyclically definable class of n-gons, or briefly a *cyclic class of n-gons*, we understand the set of solutions of a cyclic system of equations. The cyclic classes are special subspaces of the vector space of n-gons.

Since a cyclic system of equations (2) is determined by its first equation, it will often be abbreviated to

$$c_0 a_1 + c_1 a_2 + \ldots + c_{n-1} a_n = o, \ldots .$$

The element $\sum c_i$ of K is called the *coefficient sum* of the cyclic system of equations.

Every n-tuple $(c_0, c_1, \ldots, c_{n-1})$ of elements of K defines a cyclic system of equations (2) and with it a cyclic class; we speak of the cyclic class given by $(c_0, c_1, \ldots, c_{n-1})$. However, different n-tuples of elements of K may define the same cyclic class. For example, all n-tuples which differ only by a factor $c \neq 0$ of K define the same cyclic class. In the case $K = Q$, any cyclic class can be described by a coefficient n-tuple of integers.

The cyclic system of equations with coefficient n-tuple $(0, 0, \ldots, 0)$ describes the set $\mathfrak{A}_n$ of all n-gons. The cyclic system of equations $a_1 - a_2 = o, \ldots$ describes the set $\mathfrak{A}_{1,n}$ of trivial n-gons.[1] The set of solutions of the cyclic system of equations $a_1 = o, \ldots$ consists only of $(o, o, \ldots, o) = O$. Accordingly, $\mathfrak{A}_n, \mathfrak{A}_{1,n}$, and $\{O\}$ are cyclic classes.

EXAMPLE FOR $n = 4$ A quadrangle (a_1, a_2, a_3, a_4) is called a *parallelogram* when the side vectors of pairs of opposite sides have sum o: $(a_2 - a_1) + (a_4 - a_3) = o, \ldots$ We write this cyclic system of equations in the form

$$a_1 - a_2 + a_3 - a_4 = o, \ldots \tag{3}$$

The first equation says that the alternating sum of the vertices is o, and the other equations follow from the first. *The parallelograms form a cyclic class.*

If $a_1, a_2, a_3 \in V$, then the point $a_1 - a_2 + a_3$ is called *the fourth parallelogram point of the points* a_1, a_2, a_3 (figure 16).

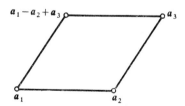

Figure 16

Exercises

1 Every cyclic class $\mathfrak{C}$ is invariant with respect to any linear transformation α of V into V: i.e. if $(a_1, a_2, \ldots, a_n) \in \mathfrak{C}$, then $(\alpha a_1, \alpha a_2, \ldots, \alpha a_n) \in \mathfrak{C}$.

[1] If $n = 1$, we have a special case: every 1-gon is trivial.

2 Is any subspace of V^n which contains, with a given n-gon $(a_1, a_2, ..., a_n)$, all n-gons which come from it by cyclic permutations a cyclic class?

3 If a cyclic class contains the n-gon $(a_1, a_2, ..., a_n)$, does it also contain the n-gon $(a_n, ..., a_2, a_1)$?

4 Every n-tuple which comes from $(c_0, c_1, ..., c_{n-1})$ by a cyclic permutation defines the same cyclic class as $(c_0, c_1, ..., c_{n-1})$.

5 If $(c_0, c_1, ..., c_{n-1})$ and $(d_0, d_1, ..., d_{n-1})$ describe the same cyclic class, then $(c_0 - d_0, c_1 - d_1, ..., c_{n-1} - d_{n-1})$ defines a cyclic class including the first cyclic class.

3

THE CENTRE-OF-GRAVITY n-GON. THE ZERO ISOBARiC CLASS

By the *centre of gravity of an n-gon* $(a_1, ..., a_n)$ we mean the point $(1/n)\sum a_i$, that is, the arithmetic mean of the vertices (the sum divided by the number). If $n = 2$, the centre of gravity is also called the mid-point.

In V^n, the vector space of n-gons, we can associate with any n-gon $(a_1, a_2, ..., a_n)$ the trivial n-gon consisting of the centre of gravity counted n times. We denote this mapping by σ:

$$\sigma: (a_1, ..., a_n) \rightarrow \left(\frac{1}{n}\sum a_i, ..., \frac{1}{n}\sum a_i\right)$$

and call the image n-gon the *centre-of-gravity n-gon* of $(a_1, a_2, ..., a_n)$. σ is a linear transformation of $\mathfrak{A}_n$ onto the class $\mathfrak{A}_{1,n}$ of trivial n-gons; for every trivial n-gon we have $\sigma(a, ..., a) = (a, ..., a)$. Therefore

$$\sigma\sigma(a_1, ..., a_n) = \sigma(a_1, ..., a_n); \tag{4}$$

i.e. the centre-of-gravity n-gon of the centre-of-gravity n-gon of $(a_1, ..., a_n)$ is the centre-of-gravity n-gon of $(a_1, ..., a_n)$. (σ is a projection; see §2.4.)

The n-gons $(a_1, ..., a_n)$ with

$$\sigma(a_1, ..., a_n) = (o, ..., o), \tag{5}$$

thus the n-gons with O as the centre-of-gravity n-gon, form a cyclic class, for (5) is simply another way of writing the cyclic system of equations

$$(1/n)\sum a_i = o, ...$$

(in which all equations are the same). This cyclic mapping is the kernel of the linear transformation σ; it is denoted by $\overset{\circ}{\mathfrak{A}}_n$ and, expressed in V, consists of all n-gons with centre of gravity o.

Two n-gons $(a_1, ..., a_n)$ and $(b_1, ..., b_n)$ are said to be *isobaric* if they have the same centre-of-gravity n-gon, thus if

$$\sigma(a_1, ..., a_n) = \sigma(b_1, ..., b_n). \tag{6}$$

(6) is equivalent to $(1/n)\sum a_i = (1/n)\sum b_i$ and so of course to $\sum a_i = \sum b_i$. 'Isobarism'

is an equivalence relation on the set $\mathfrak{A}_n$ of all n-gons which is compatible under addition. It induces a partition of the set of all n-gons into *isobaric classes*. Every n-gon is isobaric to its centre-of-gravity n-gon (because of (4)); two trivial n-gons are isobaric only if they are equal. Therefore, every isobaric class contains precisely one trivial n-gon, the common centre-of-gravity n-gon of all n-gons of the isobaric class; the trivial n-gons are a system of representatives of the isobaric classes.

$\mathring{\mathfrak{A}}_n$ is an isobaric class, the *zero isobaric class*. (Every isobaric class is a residue class $\mathring{\mathfrak{A}}_n + (a, a, ..., a)$. If $a \neq o$, this residue class is not a subspace of V^n, and so certainly not a cyclic class. $\mathring{\mathfrak{A}}_n$ is the only cyclic class which is an isobaric class.)

4
TWO TYPES OF CYCLIC CLASSES

Every cyclic class contains the trivial n-gon $O = (o, o, ..., o)$.

(i) *A cyclic class which contains a trivial n-gon $\neq O$ contains all trivial n-gons.*

PROOF Let a cyclic class be given as the solution space of a cyclic system of equations with coefficients $c_0, c_1, ..., c_{n-1} \in K$. A trivial n-gon $(a, a, ..., a)$ belongs to the cyclic class if and only if

$$c_0 a + c_1 a + ... + c_{n-1} a = o. \tag{7}$$

If there exists an $a \neq o$ such that (7) holds, then $\sum c_i = 0$. And if $\sum c_i = 0$, then (7) holds for all a.

(ii) *The centre-of-gravity n-gon of any n-gon in a cyclic class belongs to the same cyclic class.*

PROOF The centre-of-gravity n-gon of an n-gon $(a_1, a_2, ..., a_n)$ is the arithmetic mean of $(a_1, a_2, ..., a_n)$ and the n-gons which arise from it by cyclic permutations.

By (i) there are two types of cyclic classes:

I Cyclic classes which contain the class $\mathfrak{A}_{1,n}$ of trivial n-gons. If such a cyclic class, which contains all trivial n-gons $(a, a, ..., a)$, contains an n-gon $(a_1, a_2, ..., a_n)$, then it contains all n-gons $(a_1, a_2, ..., a_n) + (a, a, ..., a) = (a_1 + a, a_2 + a, ..., a_n + a)$. Since, in V, we can consider this addition of a trivial n-gon to a given n-gon $(a_1, a_2, ..., a_n)$ as a translation of $(a_1, a_2, ..., a_n)$, such a cyclic class is said to be invariant under translation in V. We call these cyclic classes *free cyclic classes*.

II Cyclic classes which contain only one trivial n-gon, the zero n-gon. By (ii)

B

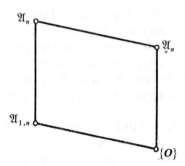

Figure 17

every such cyclic class is contained in the zero isobaric class $\overset{\circ}{\mathfrak{A}}_n$. We call these cyclic classes *zero-point classes*.

Summarizing, we have:

THEOREM 1   *There are two types of cyclic classes: the free cyclic classes, i.e. those containing the trivial n-gons; and the zero-point classes, i.e. those contained in the zero isobaric class. The free cyclic classes are the solution spaces of cyclic systems of equations with coefficient sum zero; the zero-point classes are the solution spaces of cyclic systems of equations with coefficient sum different from zero.*

$\mathfrak{A}_n$ is the largest free cyclic class, and $\mathfrak{A}_{1,n}$ the smallest. $\overset{\circ}{\mathfrak{A}}_n$ is the largest zero-point class, and $\{O\}$ the smallest. In figure 17, which demonstrates the inclusion relationships between the four 'basic classes' (a sloping line means the lower class is contained in the upper), the free cyclic classes should be thought of as being between $\mathfrak{A}_{1,n}$ and $\mathfrak{A}_n$, and the zero-point classes as being between $\{O\}$ and $\overset{\circ}{\mathfrak{A}}_n$.

EXAMPLE FOR $n = 4$ The parallelograms form a free cyclic class. The parallelograms with centre of gravity o form a zero-point class described by the cyclic system of equations $a_1 + a_3 = o, \dots$.

For every n-gon A, $A - \sigma A$ is an n-gon belonging to the zero isobaric class, since, by (4), $\sigma(A - \sigma A) = \sigma A - \sigma\sigma A = \sigma A - \sigma A = O$. Since the centre-of-gravity n-gon σA is a trivial n-gon, we can say, in V, that $A - \sigma A$ comes from A by a translation which translates A so that its centre of gravity becomes the zero point.

If A belongs to a cyclic class, then by (ii) the centre-of-gravity n-gon (and so also $A - \sigma A$) belongs to the same cyclic class. Somewhat later we shall show that under the mapping

$$A \to A - \sigma A$$

every free cyclic class is mapped onto a zero-point class contained in it, as is to be expected; and the free cyclic classes are mapped one–one onto the zero-point classes.

For the moment we shall refer only to the two types of cyclic classes. As far as cyclic classes are concerned, nothing geometrically essential would be lost if we limited ourselves to classes of one type. In our examples we shall favour the free cyclic classes.

Exercises

1 If $n \neq 1$, the four basic classes are distinct. If $n = 2$, they are the only cyclic classes.
2 If a free cyclic class $\mathfrak{C}$ is given by an n-tuple $(c_0, c_1, ..., c_{n-1})$ with $\Sigma c_i = 0$, then the n-tuple $(c_0 + 1/n, c_1 + 1/n, ..., c_{n-1} + 1/n)$ defines the class of n-gons in $\mathfrak{C}$ which have centre of gravity o.
3 If a zero-point class is defined by $(c_0, c_1, ..., c_{n-1})$, then it is also defined by all n-tuples $(c_0 - c, c_1 - c, ..., c_{n-1} - c)$ with $c \neq (1/n) \Sigma c_i$.

5
PERIODIC CLASSES

Let d be a divisor of n and $n = d\bar{d}$.

The cyclic class defined by the cyclic system of equations

$$a_1 = a_{d+1}, ... \tag{8}$$

consists of the n-gons

$$(a_1, ..., a_d, a_1, ..., a_d, ..., a_1, ..., a_d).$$

They are periodic with length (or period) d, and the number of these periods is $\bar{d}$. We denote the cyclic class defined by (8) by $\mathfrak{A}_{d,\bar{d}}$ since the n-gons of this class may be considered, in V, as d-gons counted $\bar{d}$ times (but they are actually $d \cdot \bar{d}$-gons). $\mathfrak{A}_{d,\bar{d}}$ is a free cyclic class.

For every divisor d of n there exists a class $\mathfrak{A}_{d,\bar{d}}$, called a *periodic class*. Extreme cases are $\mathfrak{A}_{n,1}$, which is the class $\mathfrak{A}_n$ of all n-gons, and $\mathfrak{A}_{1,n}$, the class of trivial n-gons.

In this connection, we introduce another geometric notation. Write the vertices of an n-gon $(a_1, a_2, ..., a_n)$ in a *vertex scheme modulo d*:

a_1	a_{d+1}	$\cdots$	a_{n-d+1}
a_2	a_{d+2}	$\cdots$	a_{n-d+2}
	$\cdots\cdots\cdots$		
a_d	a_{2d}	$\cdots$	a_n

The rows are $\bar{d}$-gons and are called the *omitting sub-$\bar{d}$-gons of* $(a_1, a_2, ..., a_n)$ (or sometimes the $(d-1)$-fold omitting subpolygons of $(a_1, a_2, ..., a_n)$). Several

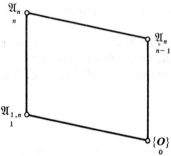

Figure 18

times we shall make use of the possibility of defining cyclic classes of n-gons by imposing conditions on the omitting subpolygons. The class $\mathfrak{A}_{d,\bar{d}}$ consists of all n-gons whose omitting sub-$\bar{d}$-gons are trivial.

6

DEGREE OF FREEDOM OF A CYCLIC CLASS

The cyclic quadrangular class of parallelograms (defined by the cyclic system of equations $a_1 - a_2 + a_3 - a_4 = o$) has the property that for any arbitrarily chosen points a_1, a_2, a_3 there is exactly one point a_4 such that (a_1, a_2, a_3, a_4) is a parallelogram. We say therefore that the cyclic class of parallelograms has degree of freedom three, and define generally: a cyclic class $\mathfrak{C}$ of n-gons has *degree (of freedom) f* if f is the maximum number of arbitrary points $a_1, a_2, ..., a_f$ which can be extended to an n-gon $(a_1, a_2, ..., a_f, a_{f+1}, ..., a_n)$ of $\mathfrak{C}$.

The degree of the cyclic class $\mathfrak{C}$ is denoted deg $\mathfrak{C}$. A few examples are shown in figure 18, where the numbers under the cyclic classes give the degrees. For periodic classes deg $\mathfrak{A}_{d,\bar{d}} = d$ (repeated countings do not alter the degree.)

Consider now the cyclic system of equations with coefficient n-tuple $(c_0, c_1, ..., c_{n-1})$. The method of writing (2) is not customary for a system of linear equations. If we write it in the standard form

$$c_0 a_1 \quad + c_1 a_2 + ... + c_{n-1} a_n = o,$$
$$c_{n-1} a_1 + c_0 a_2 + ... + c_{n-2} a_n = o,$$
$$\cdots\cdots\cdots$$
$$c_1 a_1 \quad + c_2 a_2 + ... + c_0 a_n \quad = o,$$

then we see that

$$M(c_0, c_1, ..., c_{n-1}) = \begin{bmatrix} c_0 & c_1 & \cdots & c_{n-1} \\ c_{n-1} & c_0 & \cdots & c_{n-2} \\ & & \cdots\cdots \\ c_1 & c_2 & \cdots & c_0 \end{bmatrix}$$

is the coefficient matrix of the cyclic system of equations. The elements in each row are the same but shifted, from row to row, one position to the right. Such matrices are called *cyclic matrices*.

Suppose now that the first r rows of the matrix $M(c_0, c_1, ..., c_n)$ are linearly independent, and that $(r+1)$th row is a linear combination of the previous rows. Since every row is obtained from the preceding row by shifting one position to the right, every row can be represented as a linear combination of the r previous rows. Thus r consecutive rows form a maximal system of linearly independent rows, and so the rank of the matrix is r. From the theory of systems of homogeneous linear equations, we know that at most $n-r$ consecutive points may be chosen arbitrarily and then extended uniquely to a solution of the given system of equations. Thus we have

THEOREM 2   *The degree of freedom of the cyclic class given by* $(c_0, c_1, ..., c_{n-1})$ *is*

$n - \text{rank } M(c_0, c_1, ..., c_{n-1})$.

Exercise

Let dim $V = m$. Then dim $V^n = nm$ and, if $\mathfrak{C} \subseteq V^n$ is a cyclic class with deg $\mathfrak{C} = f$, dim $\mathfrak{C} = fm$.

7
DIMENSION OF AN n-GON

We can associate a dimension to every n-gon by considering it as a set of points in V. Thus for an n-gon $(a_1, a_2, ..., a_n)$ with centre of gravity o, it is meaningful to define

dim $(a_1, a_2, ..., a_n) = $ dim $\langle a_1, a_2, ..., a_n \rangle$

where $\langle a_1, a_2, ..., a_n \rangle$ denotes the subspace of V generated by $a_1, a_2, ..., a_n$. Since $(1/n)\sum a_i = o$, this subspace has dimension at most $n-1$. For an arbitrary n-gon A, let

dim $A = $ dim $(A - \sigma A)$

so that n-gons which may be transformed into each other by a translation have the same dimension. An n-gon is then at most $(n-1)$-dimensional, every trivial n-gon has dimension 0, and triangles and parallelograms are at most 2-dimensional.

The dimension of an n-gon is the dimension of the smallest linear space in V which contains the n-gon. [The linear spaces of U are said to be the residue classes $T + a$, where T is a subspace of V, and dim $(T + a)$ is set equal to dim T.]

REMARK Let V be at least n-dimensional. For any cyclic class consider the maximum dimension which an n-gon of the class can have. This number is the same for a free class and the zero-point class associated with it:[2] it is equal to the degree of the zero-point class, whereas the degree of the free class is one greater.

8
EXAMPLES OF CYCLIC CLASSES

We shall give examples of cyclic classes for special values of n, but limit ourselves to free cyclic classes. We are particularly interested in the inclusion relationships for cyclic classes, and shall express these in diagrams: if two classes are joined in a diagram by a sloping line, the lower is contained in the upper. The numbers under the cyclic classes give the class degree. For example, see figures 19 and 20.

For a given n our assertions are always to be understood as follows: the given cyclic class exists over any field K whose characteristic does not divide n, in particular, over the field of rational numbers. Later we shall concern ourselves with the question of completeness. There can be further cyclic classes, at least over special fields.

(a) $n = p$ (prime). Up to now we know only the free cyclic classes $\mathfrak{A}_p$ and $\mathfrak{A}_{1,p}$; at any rate there cannot be any other periodic classes of p-gons, since p as a prime number has only the trivial divisors p and 1.

(b) $n = 4$. There exist three periodic quadrangular classes: $\mathfrak{A}_4$, $\mathfrak{A}_{2,2}$ (the class of digons, counted twice), and $\mathfrak{A}_{1,4}$. Further, there is the cyclic class of parallelograms.

(c) $n = 2m$. The $2m$-parallelograms were mentioned in the introduction. We

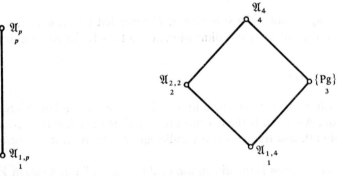

Figure 19   Figure 20

2 This concept was hinted at in §4 and will be made precise in §3.2.

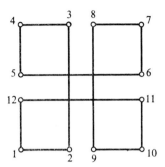

Figure 21

define the *2m-parallelograms* by the fact that the side vectors of pairs of opposite sides have sum o; that is, by the cyclic system of equations

$$a_1 - a_2 + a_{m+1} - a_{m+2} = o, \ldots.$$

An equivalent requirement is: pairs of opposite vertices have a common midpoint,

$$\tfrac{1}{2}(a_1 + a_{m+1}) = \tfrac{1}{2}(a_2 + a_{m+2}), \ldots.$$

The class of $2m$-parallelograms has degree $m+1$. Figure 21 shows a special 12-parallelogram.

The 4-parallelograms are also the quadrangles with alternating vertex sum o. Generally the cyclic system of equations

$$a_1 - a_2 + a_3 - a_4 + \ldots + a_{2m-1} - a_{2m} = o, \ldots,$$

in which all further equations follow from the first, characterizes the set of *2m-gons with alternating vertex sum* o as a cyclic class of degree $2m-1$. This property, 'the alternating sum of the vertices is o,' will often be abbreviated to ASO, and we shall speak of the ASO *class* of $2m$-gons.

THEOREM 3   *The set of 2m-parallelograms and the set of 2m-gons with alternating vertex sum* o *are cyclic classes of 2m-gons.*

When $n = 4$ these two cyclic classes coincide. This is the case only for $n = 4$, as can be seen by comparing the degrees: $m+1 = 2m-1$ implies $m = 2$ and so $n = 4$.

When $n = 4m$ we have an inclusion:

THEOREM 4   *Every 4m-parallelogram has alternating vertex sum* o.

PROOF   If $(a_1, a_2, \ldots, a_8)$ is an 8-parallelogram, then (a_1, a_2, a_5, a_6) and (a_3, a_4, a_7, a_8) are parallelograms the alternating sums of whose vertices are already o. The general result may be obtained similarly.

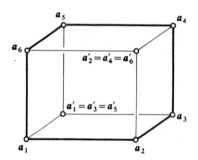

Figure 22   6-parallelogram

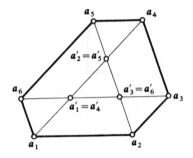

Figure 23   ASO 6-gon

(d) $n = 6$. There are four periodic classes: $\mathfrak{A}_6$, $\mathfrak{A}_{3,2}$ (class of doubly counted triangles), $\mathfrak{A}_{2,3}$ (class of triply counted digons), $\mathfrak{A}_{1,6}$, and, by theorem 3, the class of 6-parallelograms (figure 22) and the class of 6-gons with alternating vertex sum o (figure 23).

From the introduction we know the prisms as special 6-gons. The vertices of any 6-gon $(a_1, a_2, ..., a_6)$ may be arranged in the scheme

$$a_1 \quad a_3 \quad a_5$$

$$a_4 \quad a_6 \quad a_2$$

where the indices on the horizontal increase by 2's and on the vertical by 3's. We define as prisms those 6-gons for which there exists a translation in V which maps the points of the first row into the points of the second row. The cyclic system of equations

$$a_1 - a_4 = a_3 - a_6, ...$$

exhibits the set of prisms as a cyclic class of 6-gons (see figure 24).

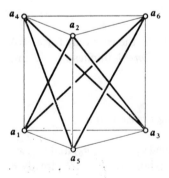

Figure 24   Prism

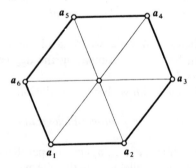

Figure 25   Affinely regular 6-gon

The *affinely regular 6-gons* are defined here as 6-gons in which any three consecutive vertices have the same fourth parallelogram point (see figure 25). The set $\mathfrak{R}_6$ of affinely regular 6-gons is a cyclic class, characterized by the cyclic system of equations

$$a_1 - a_2 + a_3 = a_2 - a_3 + a_4, \ldots.$$

Figure 26 demonstrates the inclusions which exist among these eight free cyclic hexagonal classes. If two lines slope up from one class, it can always be checked whether this class is the complete intersection of the two upper classes or not. As an example, let us prove:

THEOREM 5   *The affinely regular 6-gons are precisely the 6-parallelograms with alternating sum o.*

PROOF   For the proof we shall use characterizations of the ASO 6-gons, the 6-parallelograms, and the affinely regular 6-gons, involving the fourth parallelogram point $a_i' = a_i - a_{i+1} + a_{i+2}$ of any three consecutive points a_i, a_{i+1}, a_{i+2}.
The ASO 6-gons are the 6-gons with

$$a_1' = a_4', \qquad a_2' = a_5', \qquad a_3' = a_6';$$

the 6-parallelograms are the 6-gons with

$$a_1' = a_3' = a_5', \qquad a_2' = a_4' = a_6';$$

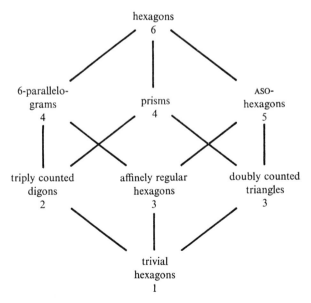

Figure 26

LIBRARY OF MOUNT ST. MARY'S COLLEGE EMMITSBURG, MARYLAND

and the affinely regular 6-gons are defined by

$$a_1' = a_2' = a_3' = a_4' = a_5' = a_6'.$$

The validity of the theorem is now obvious.

Another property of class degree is worth mentioning: in the diagram of the eight hexagonal classes (figure 26) there are three directions; the difference in degree along the lines of each direction is a constant, and the sum of these three differences is equal to the difference in degree of the largest and smallest classes of the diagram.

The first really interesting case of our theory is $n = 6$, which is recommended to the reader as a main example.

(e) $n = 8$. In addition to the four periodic classes $\mathfrak{A}_8$, $\mathfrak{A}_{4,2}$ (class of doubly counted quadrangles), $\mathfrak{A}_{2,4}$ (class of digons counted four times), and $\mathfrak{A}_{1,8}$, we have also, by theorem 3, the 8-parallelograms and the class of ASO 8-gons. By theorem 4, the latter class includes the former.

By counting twice the 4-gons from one of the four cyclic classes of 4-gons, we obtain a cyclic class of 8-gons. Thus the 4-parallelograms, doubly counted, form a cyclic class of 8-gons with defining system of equations $a_1 - a_2 + a_3 - a_4 = 0, \dots$. In contrast to this class we have the set of 8-gons in which the two omitting subpolygons are parallelograms (figure 27). This set is also a cyclic class with defining system of equations $a_1 - a_3 + a_5 - a_7 = 0, \dots$. In figure 28 it should be entered in the unmarked position.

The diagram of these eight 8-gonal classes is remarkably different from the diagram of the eight 6-gonal classes. The periodic classes form a chain, there are no prisms or affinely regular octagons, and the classes of doubly counted 4-gons form a subdiagram corresponding to the diagram of the 4-gonal classes.

(f) $n = 10$. The eight cyclic classes which are analogous to the eight 6-gonal classes are easily given.

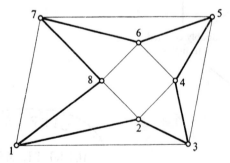

Figure 27

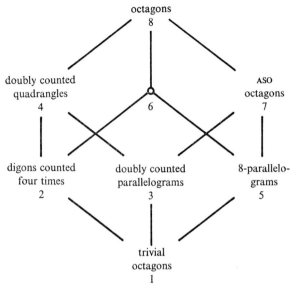

octagons
8

doubly counted
quadrangles
4

ASO
octagons
7

6

digons counted
four times
2

doubly counted
parallelograms
3

8-parallelo-
grams
5

trivial
octagons
1

Figure 28

Instead of the class of affinely regular 6-gons, we have the class defined by

$$a_1 - a_2 + a_3 - a_4 + a_5 = a_2 - a_3 + a_4 - a_5 + a_6, \ldots.$$

The regular 10-gons of the Euclidean plane and their affine images satisfy this system of equations (proof?); nevertheless figure 29 shows a 10-gon of this class which contradicts fundamentally the idea of an affinely regular 10-gon. The cyclic class has degree 5, and so contains 4-dimensional 10-gons (as long as the dimension of V is not less than 4).

A common property of 6, 8, 10 is that each has four divisors.

(g) $n = 12$. 12 has six divisors, and it can be seen that there are easily more than eight free classes.

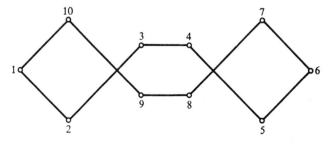

Figure 29

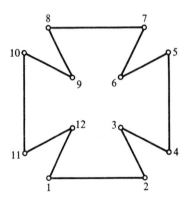

Figure 30

Let 1–6 be the periodic classes of 12-gons corresponding to the divisors $d = 12, 6, 4, 3, 2, 1$.

7–11, the quadrangles from each of the four cyclic classes of 4-gons, counted three times, form a cyclic class of 12-gons. Thus the triply counted 4-parallelograms form a cyclic class of 12-gons defined by the coefficient 12-tuple $(1, -1, 1, -1, 0, ..., 0)$. Analogously, the 6-gons from one of the eight cyclic 6-gonal classes, doubly counted, form a cyclic 12-gonal class; thus we have the ASO 6-gons, 6-parallelograms, prisms, and affinely regular 6-gons, all counted twice. If $(c_0, c_1, ..., c_5)$ is the coefficient 6-tuple by which the 6-gonal class has been defined,[3] then the corresponding 12-gonal class can be defined by the 12-tuple $(c_0, c_1, ..., c_5, 0, ..., 0)$.

12–16 are further cyclic 12-gonal classes obtained by imposing conditions on the omitting subpolygons. The 12-gons in which the three omitting subquadrangles are parallelograms form a cyclic class which can be defined by the 12-tuple $(1, 0, 0, -1, 0, 0, 1, 0, 0, -1, 0, 0)$. Analogously, the 12-gons in which the two omitting subhexagons belong to the same 6-gonal class (i.e. both are ASO 6-gons, 6-parallelograms, prisms or affinely regular 6-gons) form cyclic classes. If $(c_0, c_1, ..., c_5)$ is the defining coefficient 6-tuple of the 6-gonal class, then $(c_0, 0, c_1, 0, ..., c_5, 0)$ is a defining 12-tuple for the corresponding 12-gonal class.

The Cross Formée (figure 30) is a 12-gon in which the omitting subhexagons are affinely regular.

17–20 are classes formed by imposing conditions of isobarity (or equality of vertex sums) upon the omitting sub-d-gons. For $d = 2, 3, 4, 6$ we obtain four cyclic classes of 12-gons. If the omitting sub-2-gons (pairs of opposite vertices)

---

3 These 6-tuples were $(1, -1, 1, -1, 1, -1)$, $(1, -1, 0, 1, -1, 0)$, $(1, 0, -1, -1, 0, 1)$, $(1, -2, 2, -1, 0, 0)$.

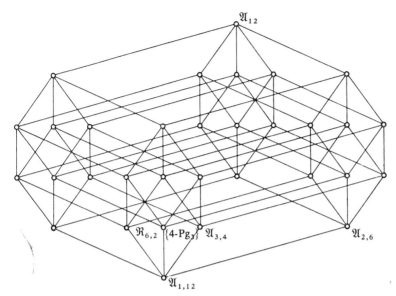

Figure 31   32 free cyclic 12-gonal classes. $\mathfrak{R}_{6,2}$: class of doubly counted affinely regular hexagons; $\{4\text{-Pg}_3\}$: class of trebly counted parallelograms.

are isobaric, then we get the 12-parallelograms; if the omitting sub-6-gons are isobaric, we get the ASO 12-gons.

In the Cross Formée the omitting, affinely regular subhexagons are isobaric, and in general for every d ($d = 2, 3, 4, 6$) the omitting sub-d-gons are isobaric.

21: the vertices of any 12-gon ($a_1, a_2, ..., a_{12}$) can be arranged in the scheme

$$a_1 \quad a_4 \quad a_7 \quad a_{10}$$

$$a_5 \quad a_8 \quad a_{11} \quad a_2$$

$$a_9 \quad a_{12} \quad a_3 \quad a_6$$

where the indices on the horizontals increase by 3's and those on the verticals by 4's. If we assume that for any two rows of the scheme there exists a translation in V which maps the points of one row onto the points of the other, we obtain the (3, 4)-prisms (analogous to the (2, 3)-prisms of 6-gons). Also, any two columns of the scheme can be mapped onto each other by a translation in V. The cyclic system of equations

$$a_1 - a_5 = a_4 - a_8, ...$$

exhibits the set of (3, 4)-prisms as a cyclic class of 12-gons.

We leave it to the reader to determine other free cyclic classes for $n = 12$ and to enter them in figure 31.

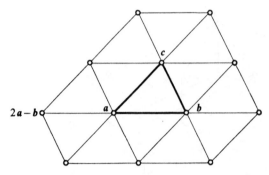

Figure 32

Exercises

1 Reflect each vertex of a triangle *a*, *b*, *c* in *V* in the two other vertices (figure 32). The hexagon thus arising $(2a-b, 2a-c, 2b-c, 2b-a, 2c-a, 2c-b)$ has alternating sum *o*.

2 Let *a*, *b*, *c*, *p* ∈ *V*. If *p* is reflected first in *a*, then in *b*, then in *c*, then again in *a*, *b*, and *c*, the figure closes, and we obtain a prism (figure 33).

3 Let *a*, *b*, *c*, *p* ∈ *V*. Form the fourth parallelogram points $a-p+b$, $b-p+c$, $c-p+a$. Then $A = (a, a-p+b, b, b-p+c, c, c-p+a)$ is a 6-parallelogram. Can every 6-parallelogram be so represented? The vertices of the hexagon $\frac{1}{2}(A+P)$ with $P = (p, p, ..., p)$ are the mid-points of the edges of the 'tetrahedron' *a*, *b*, *c*, *p*. Moreover, $\frac{1}{2}(A+P)$ is a 6-parallelogram (figure 34). The three pairs of mid-points of opposite edges of a tetrahedron have a common mid-point, the centre of gravity of the tetrahedron. What specializations appear when *p* is the centre of gravity of *a*, *b*, *c*?

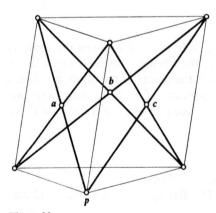

Figure 33

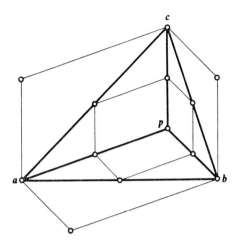

Figure 34

4 Let $n = d\bar{d}$ and $\mathfrak{C}_d$ be a cyclic d-gonal class. Then the n-gons in which all the omitting sub-d-gons (the rows of the vertex scheme modulo d) belong to the class $\mathfrak{C}_d$ form a cyclic n-gonal class. A defining n-tuple for this n-gonal class may be obtained from a defining d-tuple $(c_1, c_2, \ldots, c_d)$ by inserting $\bar{d}-1$ zeros after each c_i.

5 Let $n = d\bar{d}$ be a factorization of n into relatively prime factors. Then the cyclic system of equations $a_1 - a_{1+d} - a_{1+\bar{d}} + a_{1+d+\bar{d}} = o,\ \ldots$ defines a cyclic n-gonal class. The n-gons of this class are called $(d, \bar{d})$-*prisms*. A $(d, \bar{d})$-prism consists of $\bar{d}$ d-gons which are congruent under translation, and also of d $\bar{d}$-gons congruent under translation. The $(2, 3)$-prisms are the prisms in the class of hexagons. Every n-gon is an $(n, 1)$-prism. For which n do there exist no further classes of prisms?

6 Let $n = d\bar{d}$. If $\mathfrak{C}_d$ is a cyclic d-gonal class, then the set $\mathfrak{C}_{d,\bar{d}}$ of all n-gons which arise from the d-gons of $\mathfrak{C}_d$ by $\bar{d}$ countings is a cyclic n-gonal class. That is, the set of n-gons $(a_1, \ldots, a_d, a_1, \ldots, a_d, \ldots, a_1, \ldots, a_d)$ with $(a_1, \ldots, a_d) \in \mathfrak{C}_d$ is a cyclic n-gonal class. But in general the following does not hold: if $(c_0, c_1, \ldots, c_{d-1})$ is a defining d-tuple for $\mathfrak{C}_d$, then the n-tuple $(c_0, c_1, \ldots, c_{d-1}, 0, \ldots, 0)$ is a defining n-tuple for $\mathfrak{C}_{d,\bar{d}}$.

2
Cyclic mappings of n-gons

CYCLIC MAPPINGS

The theorems in the introduction hint at a connection between certain mappings of the set of all n-gons into itself and cyclic classes. A prototype of such a mapping is the one which maps every n-gon into the n-gon formed by the mid-points of the sides, thus the mapping

$$(a_1, a_2, ..., a_n) \rightarrow (b_1, b_2, ..., b_n)$$

with $b_1 = \frac{1}{2}(a_1+a_2)$, $b_2 = \frac{1}{2}(a_2+a_3)$, ..., $b_n = \frac{1}{2}(a_n+a_1)$. The defining system of equations of this mapping (when char $K \neq 2$) obviously has a cyclic character.

We say that a mapping of the set $\mathfrak{A}_n$ of all n-gons into itself is *cyclic* if there exist elements $c_0, c_1, ..., c_{n-1} \in K$ such that for any n-gon $(a_1, a_2, ..., a_n)$ and its image n-gon $(b_1, b_2, ..., b_n)$ we have

$$\begin{aligned}
b_1 &= c_0 a_1 + c_1 a_2 + ... + c_{n-1} a_n, \\
b_2 &= c_0 a_2 + c_1 a_3 + ... + c_{n-1} a_1, \\
&\quad\cdots\cdots\cdots\cdots \\
b_n &= c_0 a_n + c_1 a_1 + ... + c_{n-1} a_{n-1}.
\end{aligned} \qquad (1)$$

The mapping given above as an example is cyclic in this sense.

Every n-tuple $(c_0, c_1, ..., c_{n-1})$ of elements of K not only determines a cyclic class of n-gons as the solution space of a homogeneous cyclic system of equations (§1.2), but also defines a cyclic mapping of the set of n-gons into itself. The cyclic class given by $(c_0, c_1, ..., c_{n-1})$ is obviously the kernel of the cyclic mapping given by $(c_0, c_1, ..., c_{n-1})$; i.e. the set of n-gons which is mapped onto the zero n-gon $(o, o, ..., o)$. Therefore

THEOREM 1 *The cyclic classes are the kernels of the cyclic mappings.*

This connection between cyclic classes and cyclic mappings is the foundation for the further study of cyclic classes.

Different n-tuples of elements of K define different cyclic mappings:

THEOREM 2 *If two cyclic mappings defined by the n-tuples $(c_0, c_1, ..., c_{n-1})$ and $(d_0, d_1, ..., d_{n-1})$ assign to any n-gon the same image, then $c_i = d_i$ for $i = 0, 1, ..., n-1$.*

PROOF Let a be a vector distinct from o (basic assumption in §1.1). Suppose that the images of the special n-gon $(a, o, ..., o)$ under the two given cyclic mappings are the same; i.e.

$$(c_0 a, c_{n-1} a, ..., c_1 a) = (d_0 a, d_{n-1} a, ..., d_1 a).$$

Since $a \neq o$, this equation implies that $c_i = d_i$ for $i = 0, 1, ..., n-1$.

The number of cyclic mappings is thus equal to the number of n-tuples of elements from K. But there are distinct cyclic mappings which have the same kernel. Part of the aim of our theory is to determine the number of cyclic classes.

The examples in the introduction suggest the question whether every cyclic mapping maps the set of all n-gons onto a cyclic class. In this chapter and the next two we shall determine the image of the set of all n-gons for some geometrically obvious cyclic mappings. The general question will be answered in chapter 6.

2

THE ALGEBRA OF CYCLIC MAPPINGS

Before considering examples of cyclic mappings, let us lay the framework within which we shall study these mappings.

If φ is a mapping of the set $\mathfrak{A}_n$ of all n-gons into itself, then we denote the image of an n-gon A with respect to φ by φA. For the kernel of φ and the φ-image of the set $\mathfrak{A}_n$ we use the usual notations:

$$\text{Ker } \varphi := \{A : \varphi A = O\}, \quad \text{Im } \varphi := \varphi \mathfrak{A}_n = \{\varphi A : A \in \mathfrak{A}_n\}.$$

For endomorphisms $\varphi, \psi, ...$ of the vector space $\mathfrak{A}_n$ over K (linear transformations of $\mathfrak{A}_n$ into itself) let addition, multiplication by an element $c \in K$, and product be defined as usual:

$$(\varphi + \psi)A = \varphi A + \psi A, \quad (c\varphi)A = c(\varphi A), \quad (\psi\varphi)A = \psi(\varphi A).$$

The endormorphism which maps every n-gon onto the zero n-gon is denoted by 0, and the identity endomorphism by 1. The endomorphisms form, with respect to the above operations, an algebra over K denoted by $\text{End}(\mathfrak{A}_n)$; 0 and 1 are the zero and identity.

Cyclic mappings are special endomorphisms of $\mathfrak{A}_n$. The mapping

$$\zeta : (a_1, a_2, ..., a_n) \to (a_2, ..., a_n, a_1)$$

is a cyclic mapping, namely the mapping with coefficient n-tuple $(0, 1, 0, ..., 0)$. $\zeta^n = 1$, and the powers of ζ,

$$1, \zeta, \zeta^2, ..., \zeta^{n-1},\tag{2}$$

form a cyclic group of order n; they map any n-gon A onto n-gons obtained from A by cyclic permutations of the vertices. Every cyclic class is, by definition, invariant under ζ.

Let φ now be the cyclic mapping with coefficient n-tuple $(c_0, c_1, ..., c_{n-1})$. Then we can write the system of equations (1) as follows:

$$(b_1, b_2, ..., b_n)$$

$$= (c_0 a_1, c_0 a_2, ..., c_0 a_n) \qquad = c_0 1(a_1, a_2, ..., a_n)$$
$$+ (c_1 a_2, c_1 a_3, ..., c_1 a_1) \qquad + c_1 \zeta(a_1, a_2, ..., a_n)$$
$$+ .... \qquad\qquad + ...$$
$$+ (c_{n-1} a_n, c_{n-1} a_1, ..., c_{n-1} a_{n-1}) \qquad + c_{n-1} \zeta^{n-1}(a_1, a_2, ..., a_n)$$

$$= \sum_{i=0}^{n-1} c_i \zeta^i (a_1, a_2, ..., a_n).$$

Thus for any n-gon $(a_1, a_2, ..., a_n)$, $\varphi(a_1, a_2, ..., a_n)$ is equal to the last expression. Therefore

THEOREM 3 *The cyclic mapping with coefficient n-tuple $(c_0, c_1, ..., c_{n-1})$ is representable in* $\text{End}(\mathfrak{A}_n)$ *as*

$$\sum_{i=0}^{n-1} c_i \zeta^i.\tag{3}$$

Although the mapping ζ is not very interesting geometrically, we shall use it as a tool, for it enables us to write all cyclic mappings in $\text{End}(\mathfrak{A}_n)$ as linear combinations of the powers (2) of ζ, which, by theorem 2, are linearly independent. Calculation with these linear combinations is well defined by the operations in $\text{End}(\mathfrak{A}_n)$. The equations

$$\sum c_i \zeta^i + \sum d_i \zeta^i = \sum (c_i + d_i) \zeta_i, \qquad c \cdot \sum c_i \zeta^i = \sum (c c_i) \zeta^i$$

show how cyclic mappings may be added or multiplied by an element c of K. Multiplication of two linear combinations of powers of ζ is accomplished by remembering that $\zeta^n = 1$. In fact,

$$\sum d_i \zeta^i \cdot \sum c_i \zeta^i = \sum e_i \zeta^i$$

with

$$e_0 = d_0 c_0 + d_1 c_{n-1} + ... + d_{n-1} c_1,$$
$$e_1 = d_0 c_1 + d_1 c_0 + ... + d_{n-1} c_2,$$
$$..........$$
$$e_{n-1} = d_0 c_{n-1} + d_1 c_{n-2} + ... + d_{n-1} c_0.\tag{4}$$

Thus the product of two cyclic mappings is again a cyclic mapping. Since the coefficients of the product are unchanged when c_i and d_i are permuted, multiplication of cyclic mappings is commutative.

THEOREM 4   *The cyclic mappings form a commutative algebra over K (a subalgebra of $\text{End}(\mathfrak{A}_n)$); $1, \zeta, \zeta^2, \ldots, \zeta^{n-1}$ is a basis.*

The algebra of cyclic mappings, i.e. the algebra with elements (3), is the *group algebra over K of the cyclic group generated by ζ* and is denoted $K[\zeta]$. In §8.1 we shall continue with the algebraic structure of $K[\zeta]$.

Together with theorem 1, the commutativity of cyclic mappings implies that every cyclic mapping maps every cyclic class into itself:

THEOREM 5   *If ψ is a cyclic mapping and $\mathfrak{C}$ a cyclic class, then $\psi\mathfrak{C} \subseteq \mathfrak{C}$.*

PROOF   By theorem 1 $\mathfrak{C}$ is the kernel of a cyclic mapping φ: $\mathfrak{C} = \text{Ker } \varphi$. If $A \in \text{Ker } \varphi$, then $\varphi A = O$, and so a fortiori $\psi\varphi A = \psi O = O$. But since $\psi\varphi = \varphi\psi$, $\varphi\psi A = O$, i.e. $\psi A \in \text{Ker } \varphi$.

3
COEFFICIENT SUM OF A CYCLIC MAPPING

Some properties of a cyclic mapping $\varphi = \sum c_i\zeta^i$ can be obtained merely from the *coefficient sum* $s(\varphi) := \sum c_i$. The mapping

$$\varphi = \sum c_i\zeta^i \to s(\varphi) = \sum c_i \tag{5}$$

is a homomorphism of $K[\zeta]$ onto K. That $s(\psi\varphi) = s(\psi)s(\varphi)$ can be easily seen from (4).

The defining system of equations (1) of the cyclic mapping $\sum c_i\zeta^i$ implies that

$$\sum b_i = \sum c_i \sum a_i, \quad \text{and so } (1/n)\sum b_i = \sum c_i \cdot (1/n)\sum a_i. \tag{6}$$

Thus:

The cyclic mappings with coefficient sum 0 are precisely those which map every n-gon onto an n-gon with centre of gravity o, that is they map the set $\mathfrak{A}_n$ of all n-gons into the zero isobaric class $\mathfrak{A}_n$. They form the kernel of the homomorphism (5), and thus an ideal in $K[\zeta]$.

The cyclic mappings with coefficient sum 1 are precisely those which map every n-gon onto an n-gon with the same centre of gravity. These cyclic mappings are therefore more geometrically meaningful since they map isobaric classes into themselves. We call them *isobaric cyclic mappings*. Products of isobaric cyclic mappings are again isobaric cyclic mappings.

EXAMPLE   The mapping σ introduced in §1.3, which maps every n-gon onto its centre of gravity n-gon, is an isobaric cyclic mapping with coefficient n-tuple $(1/n, 1/n, \ldots, 1/n)$. Also

$$\sigma = (1/n)(1 + \zeta + \zeta^2 + \ldots + \zeta^{n-1}).$$

σ 'trivializes' every n-gon, and 'annihilates' precisely the n-gons of the zero isobaric class:

$$\text{Im } \sigma = \mathfrak{A}_{1,n}, \qquad \text{Ker } \sigma = \overset{\circ}{\mathfrak{A}}_n.$$

Theorem 1 may be made more precise: every cyclic class is the kernel of a cyclic mapping with coefficient sum 0 or of an isobaric cyclic mapping. For if φ is a cyclic mapping with $s(\varphi) \neq 0$, $[1/s(\varphi)]\varphi$ is an isobaric cyclic mapping with the same kernel. Then from theorem 1 of chapter 1 we have:

THEOREM 6   *The free cyclic classes are the kernels of the cyclic mappings with coefficient sum zero; the zero-point classes are the kernels of the isobaric cyclic mappings.*

Exercises

1 Let $A = (a_1, a_2, \ldots, a_n)$. The vertices of $(\zeta - 1)A$ are the 'side vectors' of A, $a_{i+1} - a_i$. $\text{Im}(\zeta - 1) = \overset{\circ}{\mathfrak{A}}_n$, $\text{Ker}(\zeta - 1) = \mathfrak{A}_{1,n}$.
2 The invertible cyclic mappings form a group with respect to multiplication, the group of units of $K[\zeta]$. Every invertible cyclic mapping is a one–one mapping of each cyclic class onto itself. Examples are: $c \cdot 1$ with $c \neq 0$ in K, the stretchings; $2\sigma - 1$, the reflection of each n-gon in its centre-of-gravity n-gon; ζ; $\zeta - c$, for $c^n \neq 1$. Which of these mappings are isobaric? Cyclic mappings with coefficient sum 0 are not invertible.
3 If φ is a cyclic mapping, then Fix $\varphi := \{A : \varphi A = A\}$ is a cyclic class; if, moreover, φ is isobaric, then the class is free. For example, when $d|n$, Fix $\zeta^d = \mathfrak{A}_{d,a}$; and when $n = 4$, $\text{Fix}(\zeta - \zeta^2 + \zeta^3)$ is the class of parallelograms. For every cyclic class $\mathfrak{C}$ there is a cyclic mapping φ with $\mathfrak{C} = \text{Fix } \varphi$. Study the relation Fix $\varphi = $ Fix ψ; for example, Fix $\varphi = \text{Fix}(2\varphi - 1)$ when char $K \neq 2$.

4
PROJECTIONS

First, let M be an arbitrary set with elements $a, b, \ldots$ and let $\varphi, \psi, \ldots$ be mappings of M into itself. Denote by Im φ the image of M under φ, and by Fix φ the set of fixed elements under φ:

$$\text{Im } \varphi := \{\varphi a : a \in M\}, \qquad \text{Fix } \varphi := \{a : \varphi a = a\}.$$

Then Fix $\varphi \subseteq$ Im φ, and Fix φ is the largest subset of M for which φ is the identity.

In any domain with a multiplicative operation an element φ is said to be *idempotent* if $\varphi\varphi = \varphi$. We shall call the idempotent mappings of M into itself *projections* of M. The following are three alternative ways of expressing the fact that φ is a projection of M: (1) the restriction of φ to Im φ is the identity; (2) Im φ = Fix φ; (3) Im $\varphi \subseteq$ Fix φ.

Distinct projections can have equal images. But for commutative projections it is quite remarkable that equality of images implies equality of projections:

THEOREM 7   *If φ, ψ are commutative projections with* Im φ = Im ψ, *then $\varphi = \psi$.*

PROOF   Let a be an arbitrary element of M. Since φ, ψ have equal images, and since ψ is a projection, Im $\varphi \subseteq$ Fix ψ. Thus $\varphi a \in$ Fix ψ, i.e. $\psi\varphi a = \varphi a$. Similarly we have $\varphi\psi a = \psi a$. Since $\varphi\psi = \psi\varphi$, $\varphi a = \psi a$.

A mapping of M into itself is called a *quasi-projection* of M if it maps the image of M one–one onto itself. Thus φ is a quasi-projection of M when the restriction $\varphi|_{Im\varphi}$ (restriction of φ to Im φ) is a one–one mapping of Im φ onto itself. Setting $\hat{\varphi} = \varphi|_{Im\varphi}$ we have: if φ is a quasi-projection of M, then $\hat{\varphi}^{-1}\varphi$ is a projection of M onto Im φ.

Now let $\mathfrak{A}$, $+$ be an abelian group with elements o, a, ..., and let φ, ψ, ... be endomorphisms of $\mathfrak{A}$. Im φ and Fix φ are subgroups of $\mathfrak{A}$, and there is another subgroup of $\mathfrak{A}$ determined by φ, namely the *kernel of φ*:

Ker φ: $= \{a : \varphi a = o\}$.

The endomorphisms of $\mathfrak{A}$ form a ring with respect to the addition given by $(\varphi+\psi)a = \varphi a+\psi a$ and the usual product formation $(\psi\varphi)a = \psi(\varphi a)$. The zero element of this ring is the mapping 0 which maps each element of $\mathfrak{A}$ onto o, and the identity element is the identity mapping 1. If φ is an endomorphism, so also is $1-\varphi$. Setting $1-\varphi = \varphi'$, we see that $\varphi'' = \varphi$ and so $\varphi \to 1-\varphi$ is an involution in the ring of endomorphisms of $\mathfrak{A}$. Moreover,

Fix φ = Ker$(1-\varphi)$,   Ker φ = Fix$(1-\varphi)$.

Now let φ be an idempotent endomorphism, that is, a projection of $\mathfrak{A}$. Then $1-\varphi$ is also idempotent, and φ, $1-\varphi$ are complementary to each other in the sense that

$$1 = \varphi+(1-\varphi), \qquad \varphi(1-\varphi) = (1-\varphi)\varphi = 0.$$

Moreover the equations

$$\text{Im } \varphi = \text{Ker}(1-\varphi), \qquad \text{Ker } \varphi = \text{Im}(1-\varphi), \tag{7}$$

which will often be used in the following, are valid.

THEOREM 8   *If φ is an idempotent endomorphism of an abelian group $\mathfrak{A}$, then*

$$\mathfrak{A} = \text{Im } \varphi \oplus \text{Ker } \varphi. \tag{8}$$

According to the definition of the direct sum, this equation says that Im φ and Ker φ are mutually complementary subgroups of $\mathfrak{A}$:

$$\mathfrak{A} = \text{Im } \varphi + \text{Ker } \varphi, \qquad \text{Im } \varphi \cap \text{Ker } \varphi = \{o\}.$$

PROOF OF THEOREM 8   For every $a \in \mathfrak{A}$,

$$a = 1a = \varphi a + (1 - \varphi)a \in \text{Im } \varphi + \text{Im}(1 - \varphi) = \text{Im } \varphi + \text{Ker } \varphi.$$

If $a \in \text{Im } \varphi \cap \text{Ker } \varphi$, then (because Im $\varphi = \text{Fix } \varphi$),

$$\varphi a = a \quad \text{and } \varphi a = o \quad \text{and so } a = o.$$

An endomorphism φ of $\mathfrak{A}$ is a quasi-projection if and only if

$$\text{Im } \varphi^2 = \text{Im } \varphi \quad and \quad \text{Ker } \varphi^2 = \text{Ker } \varphi. \tag{9}$$

The statement that every element of Im φ has a pre-image (under φ) in Im φ can be expressed by

$$\text{Im } \varphi \subseteq \varphi \, \text{Im } \varphi.$$

Since $\varphi \, \text{Im } \varphi = \text{Im } \varphi^2$, this can be expressed as

$$\text{Im } \varphi \subseteq \text{Im } \varphi^2.$$

The statement

$$\varphi \text{ maps Im } \varphi \text{ one--one}$$

is equivalent to

$$\varphi\varphi a = o \text{ implies } \varphi a = o,$$

and so with

$$\text{Ker } \varphi^2 \subseteq \text{Ker } \varphi.$$

Thus the statement that φ is a quasi-projection is equivalent to Im $\varphi \subseteq \text{Im } \varphi^2$ and Ker $\varphi^2 \subseteq \text{Ker } \varphi$. Since the reversed inclusions are trivial, this is equivalent to equation (9).

THEOREM 8'   *If φ is an endomorphism of an abelian group $\mathfrak{A}$, (8) holds if and only if φ is a quasi-projection of $\mathfrak{A}$.*

PROOF   The following expressions are equivalent:

for every φa there exists a φb with $\varphi a = \varphi\varphi b$;

for every a there exists a b with $\varphi a = \varphi\varphi b$, that is, with $a - \varphi b \in \text{Ker } \varphi$, that is with $a = \varphi b + (a - \varphi b)$ and $a - \varphi b \in \text{Ker } \varphi$; thus

$$\mathfrak{A} = \text{Im } \varphi + \text{Ker } \varphi.$$

Moreover, the following are equivalent statements:

$\varphi\varphi a = o$ implies $\varphi a = o$;

$\varphi a \in \text{Ker } \varphi$ implies $\varphi a = o$;

$\text{Im } \varphi \cap \text{Ker } \varphi = \{o\}$.

Among the cyclic mappings of the set $\mathfrak{A}_n$ of all n-gons into itself, there are projections; we shall call these *cyclic projections*. 0 and 1 are always cyclic projections. We can now answer, for cyclic projections, the question posed in §1 of this chapter: is the image of $\mathfrak{A}_n$ under a cyclic mapping always a cyclic class?

THEOREM 9   *If φ is a cyclic projection,* $\text{Im } \varphi$ *is a cyclic class.*

PROOF   If φ is a cyclic projection, so is $1 - \varphi$. Then $\text{Im } \varphi = \text{Ker}(1 - \varphi)$ and, by theorem 1, $\text{Ker}(1 - \varphi)$ is a cyclic class.

The coefficient sum $s(\varphi)$ of a cyclic projection φ can be only 0 or 1. For, since $\varphi \to s(\varphi)$ is a homomorphism of $K[\zeta]$ onto K, $s(\varphi)$ must be an idempotent element of K, and in a field 0 and 1 are the only idempotent elements. If $s(\varphi) = 1$, then $s(1 - \varphi) = 0$ and vice versa. Thus, together with theorem 6, we have: if φ is an isobaric cyclic projection, then $\text{Im } \varphi$ is a free cyclic class. If φ is a cyclic projection with coefficient sum zero, then $\text{Im } \varphi$ is a zero-point class.

For a cyclic projection, the image and kernel are complementary subspaces of the vector space of n-gons (theorem 8):

THEOREM 9′   *If φ is a cyclic projection, then* $\text{Im } \varphi$ *and* $\text{Ker } \varphi$ *are complementary cyclic classes. If, moreover, φ is isobaric,* $\text{Im } \varphi$ *is a free cyclic class and* $\text{Ker } \varphi$ *is a zero-point class; if φ has coefficient sum zero, then the reverse holds.*

EXAMPLE σ is an isobaric cyclic projection; $\text{Im } \sigma$ is the class $\mathfrak{A}_{1,n}$ of trivial n-gons, $\text{Ker } \sigma$ the zero isobaric class $\overset{\circ}{\mathfrak{A}}_n$ (see §1.3 and §3). $1 - \sigma$ is a cyclic projection with coefficient sum zero. Then

$$\mathfrak{A}_n = \mathfrak{A}_{1,n} \oplus \overset{\circ}{\mathfrak{A}}_n \quad \text{with} \quad \text{Im } \sigma = \mathfrak{A}_{1,n} = \text{Ker}(1 - \sigma), \quad \text{Ker } \sigma = \overset{\circ}{\mathfrak{A}}_n = \text{Im}(1 - \sigma).$$

Let A be an n-gon. Speaking geometrically, the n-gon $(1 - \sigma)A = A_0$ arises by translating A so that its centre of gravity becomes the zero point (cf. §1.4).

$A = \sigma A + (1-\sigma)A$ is the decomposition of A into its centre-of-gravity n-gon and the n-gon A_0. There is no other possible way to represent A as a sum of a trivial n-gon and an n-gon with centre of gravity o.

Exercise

An endomorphism of an abelian group is a quasi-projection precisely when it has the same image and kernel as a projection (an idempotent endomorphism). Since a projection is uniquely determined by its image and kernel, there is, for a given quasi-projection φ, precisely one projection with the same image and kernel as φ, namely $\hat{\varphi}^{-1}\varphi$.

5
EXAMPLES

The following are a few examples of isobaric cyclic mappings for $n = 4$ and $n = 6$.

(a) $n = 4$. Let us denote by κ_2 the cyclic mapping

$$(a_1, a_2, a_3, a_4) \rightarrow (b_1, b_2, b_3, b_4) \quad \text{with} \quad b_1 = \tfrac{1}{2}(a_1 + a_2), \ ...$$

which maps every quadrangle onto the quadrangle formed by the mid-points of its sides. The image quadrangle

$$\kappa_2(a_1, a_2, a_3, a_4) = (\tfrac{1}{2}(a_1 + a_2), \ ..., \ \tfrac{1}{2}(a_4 + a_1))$$

has alternating vertex sum o and is thus a parallelogram; κ_2 maps the set $\mathfrak{A}_4$ of all quadrangles into the cyclic class of parallelograms. In particular, one can verify the geometrically obvious result that κ_2 maps precisely the doubly counted digons (a_1, a_2, a_1, a_2) into trival quadrangles, that is, maps $\mathfrak{A}_{2,2}$ into $\mathfrak{A}_{1,4}$.

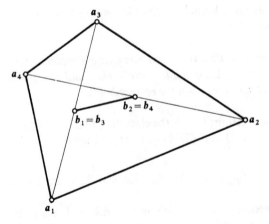

Figure 35

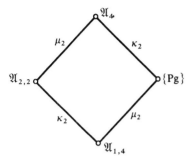

Figure 36

We look now for an isobaric cyclic mapping which maps $\mathfrak{A}_4$ into $\mathfrak{A}_{2,2}$. Obviously, this property is possessed by the cyclic mapping

$$(a_1, a_2, a_3, a_4) \to (b_1, b_2, b_3, b_4) \quad \text{with} \quad b_1 = \tfrac{1}{2}(a_1+a_3), \ldots$$

which maps every quadrangle onto the doubly counted digon formed by the mid-points of opposite vertices (figure 35). Only the parallelograms are mapped onto trivial 4-gons. Let μ_2 denote this mapping.

Then for each line segment in the diagram of the four free quadrangular classes in §1.8, we have a cyclic mapping which maps the upper class into the lower class (figure 36). The product $\kappa_2\mu_2$ maps any quadrangle onto a trivial quadrangle.

In $K[\zeta]$ we write:

$$\kappa_2 = \tfrac{1}{2}(1+\zeta), \qquad \mu_2 = \tfrac{1}{2}(1+\zeta^2)$$

and so

$$\kappa_2\mu_2 = \tfrac{1}{2}(1+\zeta)\,\tfrac{1}{2}(1+\zeta^2) = \tfrac{1}{4}(1+\zeta+\zeta^2+\zeta^3) = \sigma.$$

μ_2 is idempotent, and thus a cyclic projection, for

$$(\tfrac{1}{2}(1+\zeta^2))^2 = \tfrac{1}{4}(1+2\zeta^2+\zeta^4) = \tfrac{1}{2}(1+\zeta^2) \quad \text{since} \quad \zeta^4 = 1.$$

(b) $n = 6$. Let $\kappa_2, \kappa_3, \alpha_3$ denote the mappings $(a_1, a_2, \ldots, a_6) \to (b_1, b_2, \ldots, b_6)$ given by the following cyclic systems of equations:

$$\kappa_2 : b_1 = \tfrac{1}{2}(a_1+a_2), \ldots,$$

$$\kappa_3 : b_1 = \tfrac{1}{3}(a_1+a_2+a_3), \ldots,$$

$$\alpha_3 : b_1 = a_1-a_2+a_3, \ldots.$$

κ_2 is our prototype of a cyclic mapping and maps every hexagon onto the hexagon formed by the mid-points of its sides. With κ_2 we form the centre of gravity of any two consecutive vertices, with κ_3 we form the centre of gravity of any three consecutive vertices, and with α_3 we form the fourth parallelogram point of any three consecutive vertices (the alternating sum). The theorems in the introduction dealt with these three mappings.

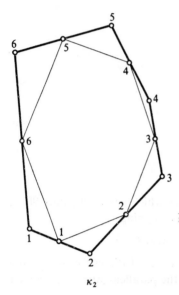

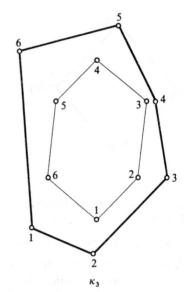

Figure 37   Figure 38

First, let us write down the following results (figures 37–9):

(i) κ_2 *maps* $\mathfrak{A}_6$ *into the class of 6-gons with alternating vertex sum* o.
(ii) κ_3 *maps* $\mathfrak{A}_6$ *into the class of 6-parallelograms.*
(iii) α_3 *maps* $\mathfrak{A}_6$ *into the class of prisms.*

These statements can be verified by elementary vector calculations:

(i) From the system of equations for κ_2 we obtain $\sum \pm b_i = o$ (where $\sum \pm b_i$ denotes the alternating sum of the b_i's).

(ii) From the system of equations for κ_3 we have

$$\tfrac{1}{2}(b_1+b_4) = \tfrac{1}{2}(b_2+b_5) = \tfrac{1}{2}(b_3+b_6) = \tfrac{1}{6}\sum a_i.$$

Thus, in the image 6-gon $\kappa_3(a_1, a_2, \ldots, a_6) = (b_1, b_2, \ldots, b_6)$, pairs of opposite vertices all have the same mid-point, namely the centre of gravity of the given 6-gon. It is also the centre of gravity of the image 6-gon.

(iii) From the system of equations for α_3 we have

$$b_1-b_4 = b_3-b_6 = b_5-b_2 = \sum \pm a_i.$$

Thus, in the image 6-gon $\alpha_3(a_1, a_2, \ldots, a_6) = (b_1, b_2, \ldots, b_6)$, the triples of vertices (b_1, b_3, b_5) and (b_4, b_6, b_2) differ by the 'vector of translation' $\sum \pm a_i$. We have $\sum \pm a_i = \tfrac{1}{3} \sum \pm b_i$.

In the diagram of the eight free hexagonal classes (figure 28), each of κ_2, κ_3, α_3 always maps an upper class into a lower class, where κ_2 operates in the NW-SE direction, κ_3 in the NE-SW direction, and α_3 in the N-S direction (figure 40).

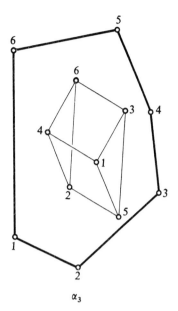

α_3

Figure 39

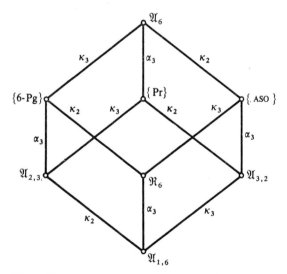

Figure 40

It is quite easy to verify without further calculation that κ_2 (for example) maps the class of 6-parallelograms into the class of affinely regular 6-gons (figure 41). Let A be a 6-parallelogram. Then $\kappa_2 A$ is, by theorem 5, a 6-parallelogram and

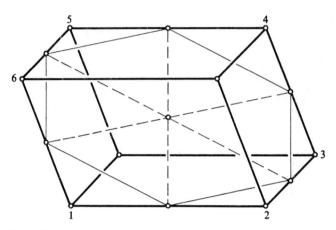

Figure 41

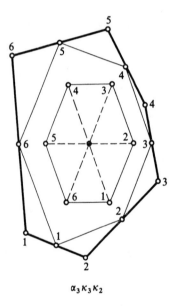

$$\alpha_3 \kappa_3 \kappa_2$$

Figure 42

has, on the other hand, alternating vertex sum *o* (by (i)). Thus by theorem 5 of chapter 1 it is affinely regular.

If we apply to an arbitrary hexagon the mappings κ_2, κ_3, α_3 in an arbitrary order, then the hexagon becomes increasingly specialized and is finally 'killed': the result is the trivial hexagon consisting merely of the centre of gravity counted six times (figure 42).

In $K[\zeta]$

$$\kappa_2 = \tfrac{1}{2}(1+\zeta), \qquad \kappa_3 = \tfrac{1}{3}(1+\zeta+\zeta^2), \qquad \alpha_3 = 1-\zeta+\zeta^2,$$

and so

$$
\begin{aligned}
\kappa_2\kappa_3\alpha_3 &= \tfrac{1}{2}(1+\zeta)\tfrac{1}{3}(1+\zeta+\zeta^2)(1-\zeta+\zeta^2) \\
&= \tfrac{1}{6}(1+\zeta+\zeta^2+\zeta^3+\zeta^4+\zeta^5) \\
&= \sigma.
\end{aligned}
$$

Exercises

1 Reflect each vertex of a pentagon $(a_1, \ldots, a_5)$ in the mid-point of its opposite side. This defines a cyclic mapping β of the set of pentagons into itself. If we apply α_3 to $\beta(a_1, \ldots, a_5)$ we obtain $(a_2, \ldots, a_5, a_1)$; β is invertible.

2 The isobaric cyclic mappings $\alpha_u: (a_1, a_2, \ldots, a_n) \to (b_1, b_2, \ldots, b_n)$ with $b_1 = a_1 - a_2 + a_3 - \ldots + a_u, \ldots$ for odd u are generalizations of α_3. Prove that $\alpha_u = 1 - \zeta + \zeta^2 - \ldots + \zeta^{u-1}$, and, if n is odd and char $K \neq 2$, that α_n, κ_2 are inverses.

6
A CYCLIC QUASI-PROJECTION

We have left open the general question as to whether every cyclic mapping maps the set of all n-gons onto a cyclic class, and have contented ourselves in the previous section with merely ascertaining whether the cyclic mappings considered there map the set of all quadrangles or hexagons into certain cyclic classes.

As an example, let us pursue the investigation further for the mapping κ_2 in the case $n = 4$.

If A is an arbitrary quadrangle, then $\kappa_2 A$, the quadrangle formed by the mid-points of the sides, is a parallelogram. If B is a given parallelogram, then we can ask: is there a quadrangle 'circumscribed' about B, i.e. a quadrangle A with $B = \kappa_2 A$? If so, what can be said about the set of circumscribed quadrangles? Are there parallelograms in the set? If so, how many?

The results are as follows: (i) $\kappa_2 A$ *is always a parallelogram.* (ii) *For every parallelogram B there are quadrangles A with $B = \kappa_2 A$.* (iii) *For every parallelogram B there is exactly one parallelogram A with $B = \kappa_2 A$.*

Together, these three statements give the following theorem:

THEOREM 10   *Let $n = 4$. The cyclic mapping κ_2 maps the set of all quadrangles onto the cyclic class of parallelograms, and the set of parallelograms one–one onto itself.*

It is geometrically evident that κ_2 is not a projection. (It can be verified that

$\frac{1}{2}(1+\zeta)$ is different from its square and also that Im $\kappa_2 \neq$ Fix κ_2.) But theorem 10 shows that κ_2 is a quasi-projection.

PROOF OF (ii) AND (iii)   Let $B = (b_1, b_2, b_3, b_4)$ be a given parallelogram; thus

$$b_1 - b_2 + b_3 - b_4 = o. \tag{10}$$

We must determine solutions of the non-homogeneous system of equations

$$b_1 = \tfrac{1}{2}(a_1 + a_2), \quad b_2 = \tfrac{1}{2}(a_2 + a_3), \quad b_3 = \tfrac{1}{2}(a_3 + a_4), \quad b_4 = \tfrac{1}{2}(a_4 + a_1). \tag{11}$$

Because of (10) the fourth equation is the alternating sum of the first three, and so dependent upon them. First, we shall determine a particular solution by choosing a vector a_1 and finding a_2, a_3, a_4 from the first, second, and third equations. For $a_1 = o$ we obtain the particular solution

$$(o, 2b_1, -2b_1 + 2b_2, 2b_1 - 2b_2 + 2b_3). \tag{12}$$

We get all the solutions of the system (11) by adding to (12) the solutions of the associated homogeneous system of equations. The solution space of this homogeneous system of equations is the cyclic class Ker κ_2; it consists of the doubly counted digons with mid-point o, i.e. the quadrangles $(c, -c, c, -c)$ for arbitrary c. Thus all the solutions of the given system of equations (11) are

$$(c, -c + 2b_1, c - 2b_1 + 2b_2, -c + 2b_1 - 2b_2 + 2b_3) \tag{13}$$

for arbitrary c.

Among these there is exactly one parallelogram. For the condition 'alternating vertex sum o' fixes c: $c = \tfrac{1}{2}(3b_1 - 2b_2 + b_3)$. Because of (10)

$$c = b_1 - b + b_4 \quad \text{where} \quad b = \tfrac{1}{4}\Sigma b_i. \tag{14}$$

The fact that the quadrangles (13) are solutions of the given system of equations is expressed geometrically as follows: to obtain a quadrangle circumscribed about the parallelogram (b_1, b_2, b_3, b_4), choose an arbitrary point c, reflect it in b_1, reflect this image in b_2, and the new image in b_3. Since (13) represents all solutions, we obtain all circumscribed quadrangles in this manner. By (14) we obtain the circumscribed parallelogram by choosing c as the fourth parallelogram point of b_1, b, b_4, where b is the centre of gravity of the given parallelogram (figure 43).

If, for a cyclic mapping, it has been determined that the set of all n-gons is mapped into a certain cyclic class, then additional consideration like this can always be made. The discussion of special cyclic classes opens a wide field of specific geometric questions. One general question, suggested by theorem 10, is the following: is every cyclic mapping a quasi-projection, i.e. does every cyclic mapping map the images of the set of all n-gons one–one onto itself? By theorem 8′ another question arises simultaneously: for every cyclic mapping are the image and kernel complementary subspaces of the vector space of n-gons?

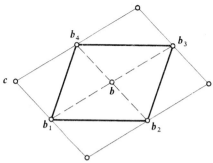

Figure 43

Exercises

1 (From G. Choquet, *Geometry in a Modern Setting* (London, 1969), p. 53.) Let char $K \neq 2$. For any n-gon B find all n-gons A for which B is the n-gon formed by the mid-points of the sides of A, i.e. $\kappa_2 A = B$. (Cf. §5, exercise 2. This shows that, if n is odd, there is precisely one 'circumscribed' n-gon for each n-gon: its vertices are the alternating sums of any n consecutive vertices of the given n-gon.)

2 Let $n = 6$. Every ASO 6-gon is the κ_2-image of exactly one ASO 6-gon. Every 6-parallelogram is the κ_3-image of exactly one 6-parallelogram. Every prism is the α_3-image of exactly one prism.

3 Let $n = 4$, char $K \neq 3$. α_3, κ_3 map $\mathfrak{A}_4$ one–one onto itself. Do A, $\alpha_3 A$ have the same edge mid-point parallelogram? If $A \in \mathfrak{A}_4$, then $\kappa_3 A = -\frac{1}{3} \zeta^{-1} A$.

4 Let $n = 5$, char $K \neq 2, 3$. κ_2, κ_3, α_3 map $\mathring{\mathfrak{A}}_5$ one–one onto itself. If $A \in \mathring{\mathfrak{A}}_5$, then $\kappa_2 \kappa_3 \alpha_3 A = \frac{1}{6} A$.

7
ISOBARIC CYCLIC PROJECTIONS FOR $n = 4$

Let $n = 4$. For each of the three periodic quadrangular classes, we already know an isobaric cyclic projection which maps the set of all quadrangles onto this class: 1, μ_2, σ respectively. Since cyclic mappings commute, there is, in general, at most one cyclic mapping with a given image (theorem 7). Thus there is no other possibility of projecting the set of all quadrangles cyclically onto the periodic quadrangular classes.

We look for a cyclic projection which projects the set of all quadrangles onto the class of parallelograms. Using theorem 10, we ask whether there is a cyclic projection with the same image as the quasi-projection κ_2. Using the notation from §4, $\hat{\kappa}_2^{-1} \kappa_2$ is the required projection.

For an arbitrary quadrangle A let $\kappa_2 A = B$ and set $\hat{\kappa}_2^{-1} \kappa_2 A = \hat{\kappa}_2^{-1} B = A^*$. Then B is the parallelogram formed by the mid-points of the sides of A, and A^* is the uniquely determined parallelogram inscribed about B (by theorem 10).

Then $\kappa_2 A^* = B = \kappa_2 A$, and hence A^* is the parallelogram with the same edge mid-point parallelogram as A.

The projection $A \to A^*$ is a cyclic mapping:

THEOREM 11   *Let $n = 4$. The mapping which maps any quadrangle onto the parallelogram with the same edge mid-point parallelogram is the cyclic projection of the set of all quadrangles onto the class of parallelograms; it is isobaric and equal to $1 - \mu_2 + \sigma$.*

PROOF   Let $A = (a_1, \ldots, a_4)$ be an arbitrary quadrangle and let B, A^* be defined as above, $A^* = (a_1{}^*, \ldots, a_4{}^*)$. Equations (13) and (14) in the previous section show what the circumscribed parallelogram looks like for an arbitrary parallelogram B. In the present situation $B = \kappa_2 A = (\frac{1}{2}(a_1 + a_2), \ldots, \frac{1}{2}(a_4 + a_1))$. Substituting this in (13) and (14), we obtain the cyclic system of equations

$$a_1{}^* = \tfrac{1}{4}(3a_1 + a_2 - a_3 + a_4), \ldots.$$

The projection $A \to A^*$ is thus the cyclic mapping with coefficient quadruple $\frac{1}{4}(3, 1, -1, 1)$. Thus in $K[\zeta]$ it is the mapping $\frac{1}{4}(3 + \zeta - \zeta^2 + \zeta^3)$. The coefficient sum is 1, and we have

$$\tfrac{1}{4}(3 + \zeta - \zeta^2 + \zeta^3) = 1 - \tfrac{1}{2}(1 + \zeta^2) + \tfrac{1}{4}(1 + \zeta + \zeta^2 + \zeta^3)$$

$$= 1 - \mu_2 + \sigma.$$

The last statement of theorem 11, written in the form

$$A^* = A - \mu_2 A + \sigma A,$$

gives a geometrical construction for the parallelogram A^* for a given quadrangle A: $a_i{}^*$ is the fourth vertex of the parallelogram formed by a_i, the mid-point of

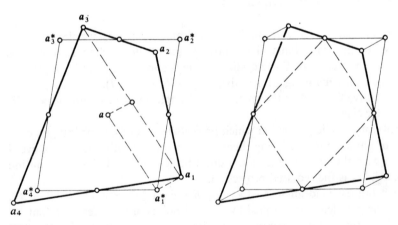

Figure 44

the 'diagonal' (a_i, a_{i+2}), and the centre of gravity of A. (In figure 44 the centre of gravity of A is the mid-point of the mid-points of the two diagonals of A.)

Thus for each of the four cyclic quadrangular classes of §1.8 we know an isobaric cyclic projection mapping the set of all quadrangles onto this class.

Again there are general questions. For arbitrary n is there a cyclic projection, for each class, projecting the set of all n-gons onto this class?

Exercises

1 Let $n = 4$. If you apply the operator '$1-$' to 1, μ_2, $1-\mu_2+\sigma$, σ then you obtain cyclic projections mapping $\mathfrak{A}_4$ onto the zero-point class $\{0\}$, the class of parallelograms with centre of gravity o, the class of doubly counted digons with mid-point o, and $\mathfrak{A}_4$ respectively. $A = A^* + (\mu_2 - \sigma)A$ is the uniquely determined decomposition of a quadrangle into a quadrangle from Im κ_2 and a quadrangle from Ker κ_2 (see theorem 8′).

2 $\text{Im}(\zeta - 1) = \mathring{\mathfrak{A}}_n = \text{Im}(1 - \sigma)$. Let $n \neq 1$. $\zeta - 1$ is a quasi-projection: if $\chi = 1/n(\zeta + 2\zeta^2 + \ldots + (n-1)\zeta^{n-1})$, then $(\zeta - 1)\chi = 1 - \sigma$ and $1 - \sigma$ is the identity map on $\mathring{\mathfrak{A}}_n$; thus χ is inverse to $\zeta - 1$. χ is, in fact, an invertible element of $K[\zeta]$: find $\psi \in K[\zeta]$ such that $\psi\chi = 1$.

3 Let $n = 4$. Determine a cyclic mapping χ with $\kappa_2\chi = 1 - \mu_2 + \sigma$. Can this equation be satisfied by a χ which is invertible?

8

CYCLIC MATRICES

If we write the system of equation (1) of the cyclic mapping with coefficient n-tuple $(c_0, c_1, \ldots, c_{n-1})$ in the usual form for a linear transformation

$$b_1 = c_0 a_1 + c_1 a_2 + \ldots + c_{n-1} a_n,$$
$$b_2 = c_{n-1} a_1 + c_0 a_2 + \ldots + c_{n-2} a_n,$$
$$\cdots\cdots\cdots$$
$$b_n = c_1 a_1 + c_2 a_2 + \ldots + c_0 a_n,$$

then we see that the coefficient matrix is the cyclic matrix $M(c_0, c_1, \ldots, c_{n-1})$ defined in §1.6.

The cyclic matrix $Z = M(0, 1, 0, \ldots, 0)$ corresponds to the cyclic mapping ζ. Z is a permutation matrix with $Z^n = E$ (E is the identity matrix); thus we have

$$M(c_0, c_1, \ldots, c_{n-1}) = \Sigma c_i Z^i.$$

The mapping

$$\Sigma c_i \zeta^i \rightarrow \Sigma c_i Z^i$$

C

is an isomorphism of the algebra $K[\zeta]$ onto the *algebra* $K[Z]$ *of* $n \times n$ *cyclic matrices over K*. Thus to the product of cyclic mappings there corresponds the multiplication of the associated cyclic matrices; it is important that this multiplication is commutative. For the K-algebra of cyclic mappings of n-gons we have a representation in the field K of degree n. The theory of cyclic $n \times n$ matrices is another interpretation of the theory of cyclic mappings of n-gons.

A cyclic class of n-gons can be characterized in an obvious way as the set of all n-gons annihilated by a cyclic matrix.

We have mentioned the obvious and interesting approach which cyclic matrices give to the study of cyclic mappings and cyclic classes here, since we favour the representation of cyclic mappings in the algebra $K[\zeta]$.

Exercises

1 Regular $n \times n$ cyclic matrices form an abelian group with respect to matrix multiplication.

2 The regular $n \times n$ diagonal matrices D which transform every cyclic $n \times n$ matrix T into cyclic matrices (i.e. those matrices for which $D^{-1}TD$ is a cyclic matrix) are precisely the diagonal matrices whose main diagonal elements are

$$c, cw, cw^2, \ldots, cw^{n-1},$$

with $c \neq 0$ and w an nth root of unity in K. Determine the matrices D over the fields of rational and real numbers, respectively (see chapter 9.5, exercise 3).

3 Let K be a field containing the nth roots of unity, and let w be a primitive root of unity. The regular matrix

$$U = (u_{i,j}) \quad \text{with} \quad u_{i,j} = w^{i \cdot j}$$

transforms every $n \times n$ cyclic matric into a matrix in diagonal form. Let T be the matrix belonging to the n-tuple $c_0, c_1, \ldots, c_{n-1}$; then the diagonal matrix $D = U^{-1}TU$ has main diagonal entries

$$d_i = c_0 + c_1 w^i + c_2 w^{2i} + \ldots + c_{n-1} w^{(n-1)i}$$

for $i = 1, 2, \ldots, n$. What conditions result for idempotent cyclic matrices?

3
Isobaric cyclic mappings

σ-KERNEL

Among cyclic mappings the isobaric mappings are distinguished geometrically in that they map every isobaric class into itself, and so operate in the individual isobaric classes. 1, σ, ζ are examples.

By definition, a cyclic mapping φ is isobaric when it carries every n-gon over into one which is isobaric to it, that is, if for every n-gon A, $\sigma\varphi A = \sigma A$, i.e. if

$$\sigma\varphi = \sigma. \tag{1}$$

Each of the following is equivalent to (1):

$$(1-\sigma)\varphi = \varphi - \sigma, \tag{2}$$

$$s(\varphi) = 1, \tag{3}$$

$$\mathfrak{A}_{1,n} \subseteq \text{Fix } \varphi. \tag{4}$$

We know from §2.3 that a cyclic mapping is isobaric precisely when its coefficient sum is 1. The equivalence of (1) and (3) also follows from the following

LEMMA *For every cyclic mapping φ, $\varphi\sigma = s(\varphi)\sigma$.*

This can be proved by referring back to the multiplication of cyclic mappings (§2.2).

As for (4), the statement $\varphi\sigma = \sigma$, which is equivalent to (1), says that for any n-gon A, $\varphi(\sigma A) = \sigma A$. A cyclic mapping is therefore isobaric if and only if it leaves fixed every centre-of-gravity n-gon, i.e. every trivial n-gon.

Now let φ be an isobaric cyclic mapping. The kernel of φ is a zero-point class (chapter 2, theorem 6). In addition to the set of n-gons which are annihilated by φ, we can consider the set of n-gons which are mapped by φ onto their centre-of-gravity n-gon (their centre of gravity counted n times). This set is called the *σ-kernel of φ*.

$$\underset{\sigma}{\text{Ker }} \varphi := \{A : \varphi A = \sigma A\} = \text{Ker}(\varphi - \sigma).$$

The σ-kernel of φ is a cyclic class (chapter 2, theorem 1); it contains all trivial n-gons (for every trivial n-gon A, $\sigma A = A$, and by (4) $\varphi A = A$, and so $\varphi A = \sigma A$) and thus is a free cyclic class. Because $s(\varphi - \sigma) = s(\varphi) - s(\sigma) = 1 - 1 = 0$, this also follows from chapter 2, theorem 6.

Exercise

Prove $\zeta\sigma = \sigma$ and use this to prove the lemma.

2

TWO TYPES OF CYCLIC CLASSES

THEOREM 1   *Every cyclic class is either the σ-kernel or the kernel of an isobaric cyclic mapping. Free cyclic classes are σ-kernels, zero-point classes are kernels, of isobaric cyclic mappings.*

PROOF   By chapter 2, theorem 6, we have to show only that the kernel of a cyclic mapping ψ with $s(\psi) = 0$ is also the σ-kernel of an isobaric cyclic mapping. Set $\psi + \sigma = \varphi$. Then $s(\varphi) = 1$ and $\operatorname{Ker} \psi = \operatorname{Ker}_{\sigma} \varphi$.

In order to describe relationships between free cyclic classes and zero-point classes, we shall prove the following for an isobaric cyclic mapping φ:

$$\operatorname{Ker}_{\sigma} \varphi \cap \operatorname{Ker} \sigma = \operatorname{Ker} \varphi, \tag{5}$$

$$\operatorname{Ker}_{\sigma} \varphi = \operatorname{Im} \sigma + \operatorname{Ker} \varphi, \tag{6}$$

$$(1 - \sigma) \operatorname{Ker}_{\sigma} \varphi = \operatorname{Ker} \varphi. \tag{7}$$

$\operatorname{Ker} \sigma$ is the zero isobaric class, and $\operatorname{Im} \sigma = \operatorname{Fix} \sigma$ is the class of trivial n-gons (see §2.3).

PROOF OF (5) AND (7)   For every n-gon A, the following three expressions are equivalent:

(a) $\varphi A = O$,
(b) $\varphi A = \sigma A = O$,
(c) there exists a B with $A = (1 - \sigma)B$ and $\varphi B = \sigma B$.

(a) implies that $\sigma \varphi A = \sigma O = O$ and so, by (1), $\sigma A = O$, and thus we have (b); (b) implies that $A = (1 - \sigma)A$ and $\varphi A = \sigma A$ which gives (c). Finally, assuming (c) and using (2), $\varphi A = \varphi(1 - \sigma)B = (\varphi - \sigma)B = \varphi B - \sigma B = O$, which gives us (a).

Equation (5) is the equivalence of (b) and (a); equation (7) is the equivalence of (c) and (a).

Equation (6) can be verified similarly, or can be proved from (5). The modular law

$$\text{if } \mathfrak{B} \subseteq \mathfrak{D}, \text{ then } (\mathfrak{B} + \mathfrak{C}) \cap \mathfrak{D} = \mathfrak{B} + (\mathfrak{C} \cap \mathfrak{D})$$

(see appendix 1) holds for subspaces of a vector space, and so for cyclic classes. Since the σ-kernel of φ is a free class, we know that $\text{Im } \sigma \subseteq \underset{\sigma}{\text{Ker }} \varphi$, and

$$\underset{\sigma}{\text{Ker }} \varphi = \mathfrak{A}_n \cap \underset{\sigma}{\text{Ker }} \varphi = (\text{Im } \sigma + \text{Ker } \sigma) \cap \underset{\sigma}{\text{Ker }} \varphi = \text{Im } \sigma + (\text{Ker } \sigma \cap \underset{\sigma}{\text{Ker }} \varphi)$$

$$= \text{Im } \sigma + \text{Ker } \varphi \quad \text{(by (5))}.$$

Because of theorem 1, and (5) and (6), the mapping

$$\underset{\sigma}{\text{Ker }} \varphi \to \text{Ker } \varphi \quad (\varphi \text{ isobaric})$$

is a one–one mapping of the set of free cyclic classes onto the set of zero-point classes. Thus a zero-point class is determined for every free cyclic class $\mathfrak{C}$; we shall call it the associated zero-point class and denote it by $\overset{\circ}{\mathfrak{C}}$. We note the following

RULE   *If φ is isobaric and $\mathfrak{C} = \underset{\sigma}{\text{Ker }} \varphi$, then $\overset{\circ}{\mathfrak{C}} = \text{Ker } \varphi$*

and collect these results into

THEOREM 2   *The mapping $\mathfrak{C} \to \overset{\circ}{\mathfrak{C}}$ which maps each free cyclic class onto the associated zero-point class is a one–one mapping of the set of free cyclic classes onto the set of zero-point classes. Moreover*

$$\mathfrak{C} \cap \mathfrak{A}_n = \overset{\circ}{\mathfrak{C}}, \qquad \mathfrak{C} = \mathfrak{A}_{1,n} + \overset{\circ}{\mathfrak{C}}, \qquad (1 - \sigma)\mathfrak{C} = \overset{\circ}{\mathfrak{C}}.$$

The most natural way of mapping an n-gon A onto an n-gon with centre of gravity o is to shift A so that its centre of gravity becomes the zero point o. This mapping is, however, the cyclic projection $1 - \sigma$ of §2.4. The last statement of theorem 2 now shows that this cyclic projection maps the set of free cyclic classes one–one onto the set of zero-point classes (see figure 45).

Theorem 2 has made precise the connection between the two types of cyclic classes. If we consider only the geometric interest of cyclic classes, it would seem that we should limit ourselves to one of the two types of cyclic classes, since the other will contribute nothing geometrically new. But a geometrically motivated limitation to the free cyclic classes would lead to a certain algebraic unevenness, as would the preference for isobaric cyclic mappings. The following section and the concluding notes contain a few hints about this. We can make do as long as we consider examples of cyclic classes and cyclic mappings, as will be done in §§4.1–2. Nevertheless, if we wish to obtain a full insight into the connections

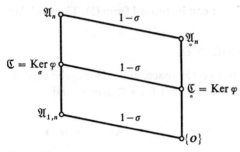

Figure 45

between cyclic classes and cyclic mappings, and formulate and prove the main theorem of our theory of n-gons, we shall have to use both concepts in their natural algebraic generality.

Figure 46 shows the inclusion diagram of the 16 cyclic hexagonal classes, which are obtained by adjoining to the eight hexagonal classes of §1.8 their associated zero-point classes (for the notation see §4.1).

REMARK 1   Let φ again be an isobaric cyclic mapping. Im φ contains the class of trivial n-gons, as can be concluded from (4) $[\mathfrak{A}_{1,n} \subseteq \text{Fix } \varphi \subseteq \text{Im } \varphi]$. If Im φ is a cyclic class, then Im φ is also a free class. The equation $\psi\text{Im } \varphi = \text{Im } \psi\varphi$, which holds for arbitrary mappings of a set into itself, shows that for $\psi = 1 - \sigma$, because of (2), $(1 - \sigma)\text{Im } \varphi = \text{Im}(\varphi - \sigma)$. Therefore, by theorem 2, we have for a (free) cyclic class $\mathfrak{C}$, the

RULE   *If φ is isobaric and $\mathfrak{C} = \text{Im } \varphi$, then $\overset{\circ}{\mathfrak{C}} = \text{Im}(\varphi - \sigma)$.*

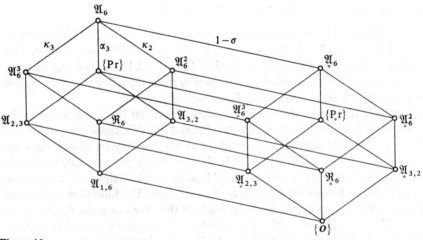

Figure 46

REMARK 2 Two subspaces $\mathfrak{B}$, $\mathfrak{C}$ of a vector space $\mathfrak{A}_n$ are by definition complementary if

$$\mathfrak{A}_n = \mathfrak{B} + \mathfrak{C}, \qquad \mathfrak{B} \cap \mathfrak{C} = \{O\}, \qquad \text{i.e. } \mathfrak{A}_n = \mathfrak{B} \oplus \mathfrak{C}.$$

Given two complementary cyclic classes, one is always a free cyclic class and the other a zero-point class. For two free cyclic classes could not be complementary since their intersection would contain at least the class of trivial n-gons, and two zero-point classes could not be complementary since their sum is at most the zero isobaric class.

Exercises

1 The set of free cyclic classes is mapped one–one onto the set of zero-point classes by the cyclic mapping $\zeta - 1$. (Cf. the exercises in §§2.3 and 2.7.)
2 If we apply the cyclic mappings κ_2, κ_3, α_3, $1 - \sigma$ one after another, in an arbitrary order, to a hexagon, the hexagon is increasingly specialized, corresponding to the sloping lines of figure 46, and finally it is mapped into O. In fact, $\kappa_2 \kappa_3 \alpha_3 (1 - \sigma) = 0$.

3
REMARKS CONCERNING ISOBARIC CYCLIC MAPPINGS

The cyclic mappings φ with $s(\varphi) = 0$ form an ideal I_0 in $K[\zeta]$, the kernel of the homomorphism $\varphi \to s(\varphi)$ of $K[\zeta]$ onto K (§2.3; by the homomorphism theorem, $K[\zeta]/I_0 \cong K$, and I_0 is a maximal ideal of $K[\zeta]$). For every $c \in K$, the set I_c of cyclic mappings φ with $s(\varphi) = c$ is a residue class of the ideal I_0. I_1 is the set of isobaric cyclic mappings, and, since $\sigma \in I_1$, we can write

$$I_1 = I_0 + \sigma. \tag{8}$$

If φ, $\psi \in I_1$, then $\varphi\psi \in I_1$ but $-\varphi$, $\varphi + \psi \notin I_1$. Nevertheless, the alternating sum of an odd number of elements of I_1 is again in I_1. The operator $1 -$ takes elements out of I_1: in fact, the mapping $\varphi \to 1 - \varphi$ interchanges I_1 and I_0.
However

$$\varphi \to 1 - \varphi + \sigma \tag{9}$$

is an involutory mapping of I_1 onto itself. If φ is an idempotent element of I_1, then $1 - \varphi + \sigma$ is also idempotent; these two isobaric cyclic projections are σ-complementary in the following sense: their product is σ, and their 'Boolean sum' (sum minus product: see §5.1) is 1.

RULE *If φ is an isobaric cyclic projection, then $1 - \varphi + \sigma$ is also an isobaric cyclic projection and*

$$\operatorname{Im}(1 - \varphi + \sigma) = \operatorname*{Ker}_{\sigma} \varphi, \qquad \operatorname*{Ker}_{\sigma}(1 - \varphi + \sigma) = \operatorname{Im} \varphi. \tag{10}$$

Equations (10) follow from $\text{Im } \varphi = \text{Ker}(1-\varphi)$, which is valid for every cyclic projection, and from the definition of σ-kernel. Both sets (10) are free cyclic classes (theorem 1).

EXAMPLE   1, σ are σ-complementary. If $n = 4$, the set of four isobaric cyclic projections in §2.7 can be divided into two pairs which are mutually σ-complementary: 1, σ and μ_2, $1-\mu_2+\sigma$. Then $\text{Im}(1-\mu_2+\sigma) = \underset{\sigma}{\text{Ker }} \mu_2$ is the class of parallelograms.

Exercise

The principal ideal (σ) generated by σ in $K[\zeta]$ is $K\sigma$ and consists of the cyclic mappings $\Sigma c_i \zeta^i$ with $c_0 = c_1 = \ldots = c_{n-1}$. Also $(1-\sigma) = I_0$ and $K[\zeta] = K\sigma \oplus I_0$.

Notes concerning the addition of n-gons

1
ADDITION OF n-GONS

Let two n-gons $A = (a_1, a_2, \ldots, a_n)$, $B = (b_1, b_2, \ldots, b_n)$ be given, which we shall interpret as point n-tuples of V.

Assume first that A and B belong to the zero isobaric class, and thus have the same centre of gravity o. Then the sum $A+B$ which was defined generally in §1.1 has a clear geometric meaning: the ith vertex of $A+B$ is the fourth vertex of the parallelogram determined by a_i, the common centre of gravity o of A and B, and b_i. $A+B$ is also an n-gon with centre of gravity o. Thus, for example, when $n = 6$, a triply counted 2-gon and a doubly counted 3-gon, both having centre of gravity o, form under addition a prism with centre of gravity o (figure 47).

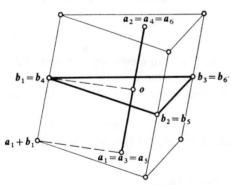

Figure 47

Suppose now that A and B are arbitrary n-gons. Then the ith vertex of $A+B$ is still the fourth vertex of the parallelogram determined by a_i, o, and b_i, but the origin o no longer has to have any geometrical relation with A and B; it can be anywhere at all. Nevertheless, the sum of, for example, a triply counted 2-gon $(a_1, a_2, a_1, a_2, a_1, a_2)$ and a doubly counted 3-gon $(b_1, b_2, b_3, b_1, b_2, b_3)$ is always a prism; for the sum hexagon $(c_1, c_2, ..., c_6)$ satisfies $c_1 - c_4 = c_3 - c_6$, ... (namely $(a_1 + b_1) - (a_2 + b_1) = (a_1 + b_3) - (a_2 + b_3)$, ...).

For the centre-of-gravity n-gons of A, B, and $A+B$ we have: $\sigma(A+B) = \sigma A + \sigma B$. Let A, B, be isobaric, and let $S = (s, s, ..., s)$ be their common centre-of-gravity n-gon: $\sigma A = \sigma B = S$. Then $\sigma(A+B) = S$ if and only if $S = O$. Thus, in general, $A+B$ does not lie in the same isobaric class as A, B. Let us form from A, B the n-gon whose ith vertex is the fourth vertex of the parallelogram formed by a_i, s, b_i, and say that it arises from A, B by *addition with respect to the centre of gravity*. It is the n-gon $A+B-S$ and lies in the same isobaric class as A and B.

The next section should clarify algebraically the formal properties of this addition with respect to the centre of gravity.

Exercise

In a Euclidean plane, after choice of an origin, the quadrangles form a group with respect to addition, and the parallelograms a subgroup. Do the squares form a subgroup?

2

DISPLACEMENT OF ADDITION IN AN ABELIAN GROUP

Let $\mathfrak{A}$, $+$ be an abelian group and $s \in \mathfrak{A}$. We can define a new operation

$$a +_s b = a + b - s \tag{1}$$

for which s is the neutral element. We obtain this operation $(a, b) \to a +_s b$ by transforming the given addition $(a, b) \to a + b$ by the translation

$$x \to x + s, \tag{2}$$

i.e. by the series of mappings

$$(a, b) \to (a - s, b - s) \to (a - s) + (b - s) \to ((a - s) + (b - s)) + s = a + b - s.$$

$\mathfrak{A}$, $+$ is mapped isomorphically onto $\mathfrak{A}$, $+_s$ by the translation (2), since $(x + s) +_s (y + s) = (x + y) + s$. Therefore, $\mathfrak{A}$, $+_s$ *is an abelian group with neutral element s.*

If $\mathfrak{U}$, $+$ is a subgroup of $\mathfrak{A}$, $+$, then (2) permutes the residue classes of $\mathfrak{U}$ and, in particular, maps $\mathfrak{U}$, $+$ isomorphically onto $\mathfrak{U} + s$, $+_s$. *Thus we can change a given residue class into a group by displacing the addition.*

If φ is an endomorphism of $\mathfrak{A}$, $+$ and if $s \in$ Fix φ, then the set of pre-images of s under φ.

$$\{c \in \mathfrak{A} : \varphi c = s\} = \operatorname{Ker} \varphi + s, \tag{3}$$

is an abelian group with respect to the operation $+_s$.

Finally let R, $+$, $\cdot$ be a commutative ring with elements a, b, ..., and let s be an idempotent element of R. Then the mapping $x \to sx$, the multiplication by s, is an idempotent endomorphism of R, with s as a fixed element and kernel $I = \{x \in R : sx = 0\}$; moreover,

$$\{c \in R : sc = s\} = I + s. \tag{4}$$

We have $(x+s)(y+s) = xy + s$ for all x, $y \in I$; therefore I, $+$, $\cdot$ is mapped isomorphically onto $I + s$, $+_s$, $\cdot$ by $x \to x + s$. The set (4) is, thus, a ring with respect to $+_s$, $\cdot$ with s as zero element.

Exercises

1 Let $\mathfrak{A}$, $+$ be a vector space over a field K and let $s \in \mathfrak{A}$. Then $\mathfrak{A}$, $+_s$ is a vector space with the given scalar multiplication $(c, a) \to ca$ $(c \in K, a \in \mathfrak{A})$ if and only if $s = o$.

2 Let R, $+$, $\cdot$ be a ring, $s \in R$, and T be a subring. $T + s, +_s, \cdot$ is a ring if and only if $s^2 = s$ and $sT = \{0\} = Ts$.

3
ADDITION OF ISOBARIC n-GONS WITH RESPECT TO THE CENTRE OF GRAVITY

An isobaric class consists of n-gons C such that $\sigma C = S$ where S is their common centre-of-gravity n-gon; it is the residue class $\mathfrak{A}_n + S$ of the zero-point class. With respect to addition,[1] defined by

$$A +_\sigma B = A + B - S$$

(addition with respect to the centre of gravity), the isobaric class is an abelian group with S as neutral element. We have $A -_\sigma B = A - B + S$.

We shall not make explicit use of this displaced addition. But it is of interest that certain sums, namely,

(i) the sums of n-gons in the zero isobaric class, and
(ii) the alternating sums of an odd number of isobaric n-gons,

coincide with the corresponding σ-sums and can therefore be interpreted geometrically in the sense of addition with respect to the centre of gravity. (For (ii) note that $A - B + C = A -_\sigma B +_\sigma C$ for isobaric A, B, C.)

1 In $A +_\sigma B$ we write σ as an abbreviation for the common centre-of-gravity n-gon $\sigma A = \sigma B$.

EXAMPLE OF (i)   Part of the aim of our theory is to show that every n-gon can be represented as a sum of n-gons from atomic cyclic classes. A cyclic n-gonal class is said to be 'atomic' if it is distinct from the zero class $\{O\}$ and contains properly no cyclic class except the zero class. Then from chapter 1, theorem 1 we have: *The class $\mathfrak{A}_{1,n}$ of trivial n-gons is an atomic cyclic class. All other atomic cyclic classes are zero-point classes.* (The class of trivial n-gons is the smallest free cyclic class and so there can be no other free cyclic classes which are atomic. To see that it is atomic, we note that if $\mathfrak{C}$ is a cyclic class contained properly in it, then $\mathfrak{C}$ is a zero-point class, and so a class of trivial n-gons with centre of gravity o, that is, the zero class.)

Therefore, every sum of n-gons from atomic cyclic classes $\neq \mathfrak{A}_{1,n}$ is a sum of type (i); addition of trivial n-gons merely implies in V a translation of the n-gons already obtained.

EXAMPLE OF (ii)   Every alternating sum of an odd number of isobaric images of a given n-gon A is a sum of type (ii); for, if φ, ψ, χ are isobaric cyclic mappings, then $(\varphi - \psi + \chi)A = \varphi A - \psi A + \chi A$ is a sum of type (ii). If in particular φ is an isobaric cyclic projection and $1 - \varphi + \sigma$ is its σ-complement, then

$$(1 - \varphi + \sigma)A = A - \varphi A + \sigma A = A -_\sigma \varphi A$$

and the ith vertex of this n-gon is the fourth vertex of the parallelogram determined by the ith vertices of A, φA, and the centre of gravity of A. For $n = 4$, this includes the construction of §2.7 for the parallelogram $(1 - \mu_2 + \sigma)A$ which has the same edge mid-point parallelogram as the quadrangle A.

Every isobaric cyclic mapping φ acts as an endomorphism on every isobaric class which has been made into an abelian group by σ-addition; for $\varphi(A +_\sigma B) = \varphi A +_\sigma \varphi B$ holds for all isobaric n-gons A, B (see chapter 3, equation (1)). With this displaced addition we can even go a step further. With respect to σ-addition defined by

$$\varphi +_\sigma \psi = \varphi + \psi - \sigma$$

and multiplication given in $K[\zeta]$, the set of isobaric cyclic mappings (the set $\{\varphi \in K[\zeta] : \sigma\varphi = \sigma\}$) is a commutative ring with σ as zero element. Thus, for example, $1 -_\sigma \varphi = 1 - \varphi + \sigma$, and for every n-gon A, $(\varphi +_\sigma \psi)A = \varphi A +_\sigma \psi A$. With operations $+_\sigma$, $\cdot$, the set of isobaric cyclic mappings is a ring of endomorphisms for every isobaric class if its n-gons are added with respect to the centre of gravity.

These are all the hints we shall give about the possibility of dealing with isobaric cyclic mappings and their σ-kernels, the free cyclic classes. We shall no longer consider as an isolated problem this particular aspect of n-gonal theory, which is independent of the choice of origin.

4
Averaging mappings

ISOBARICALLY SPLITTING n-GONS

Let us denote by $\tau(n)$ the number of divisors of n, which can be easily read from the exponents appearing in the prime factorization of n.

Let d be a divisor of n, and $n = d\bar{d}$. If we write the vertices of an n-gon $(a_1, a_2, \ldots, a_n)$ in the vertex scheme modulo d, we have

$$
\begin{array}{cccc}
a_1 & a_{d+1} & \cdots & a_{n-d+1} \\
a_2 & a_{d+2} & \cdots & a_{n-d+2} \\
& \cdots\cdots\cdots & & \\
a_d & a_{2d} & \cdots & a_n
\end{array}
\tag{1}
$$

The rows are $\bar{d}$-tuples, the omitting sub-$\bar{d}$-gons of the given n-gons (cf. §1.5). There are d of them, and we also call them the d omitting subpolygons of (1).

We can often define cyclic classes by imposing conditions on the omitting subpolygons. If for a given d we impose the condition that the d omitting subpolygons of (1) are trivial, we obtain the periodic class $\mathfrak{A}_{d,\bar{d}}$. There are $\tau(n)$ periodic classes, and we can arrange them in an inclusion diagram according to the diagrams of the divisors of n.

We say that $(a_1, a_2, \ldots, a_n)$ is d *times isobarically split* if the d omitting subpolygons of (1) are all isobaric. We see immediately that the common centre of gravity of the subpolygons is the centre of gravity of the given n-gons. The n-gons which are d times isobarically split form a free cyclic class $\mathfrak{A}_n^d$ which can be described by the cyclic system of equations

$$
(d/n)(a_1 + a_{d+1} + \ldots + a_{n-d+1}) = (1/n)\sum a_i, \ldots.
\tag{2}
$$

Extreme cases are $\mathfrak{A}_n^1 = \mathfrak{A}_n$, $\mathfrak{A}_n^n = \mathfrak{A}_{1,n}$.

Further examples, for $n = 2m$, are: $\mathfrak{A}_{2m}^2$ is the class of $2m$-gons with alternating vertex sum o, and $\mathfrak{A}_{2m}^m$ is the class of $2m$-parallelograms.

For a given n, there is a class $\mathfrak{A}_n^d$ for every divisor d, and so altogether there are $\tau(n)$ classes of this type. They can be arranged in an inclusion diagram corresponding to the diagram of the divisor d (which appears at the top of the symbol $\mathfrak{A}_n^d$).

For $n = 6$, $\mathfrak{A}_{2,3} + \mathfrak{A}_{3,2}$ is the class of prisms (cf. the notes about the addition of n-gons) and $\mathfrak{A}_6^2 \cap \mathfrak{A}_6^3$ (i.e. the set of hexagons which can be split into 2 isobaric 3-gons as well as into 3 isobaric 2-gons) is the class of affinely regular hexagons (chapter 1, theorem 5). As this example shows, the sum of two periodic classes is not in general a periodic class, and the intersection of two classes of type $\mathfrak{A}_n^d$ is not in general another class of this type. Instead there are possibilities of getting further cyclic classes by adding periodic classes or taking the intersections of classes of type $\mathfrak{A}_n^d$.

REMARK ON NOTATION   In order to avoid confusion between d and the 'complementary' divisor $\bar{d}$ of n, notice that the upper index in the symbol $\mathfrak{A}_n^d$ gives the number of isobaric subpolygons into which the n-gons of this class split, and *not* the number of vertices of the subpolygons, which is considered secondary. In the notation for this class, and also the mapping μ_d which will be defined shortly, we have used the divisor d in preference to the divisor $\bar{d}$, since this concept depends upon the vertex scheme modulo d.

Exercises

1 The cyclic class $\mathfrak{A}_n^d$ has degree $n - d + 1$.
2 Let $A = (a_1, a_2, ..., a_n)$. The vertices of $(\zeta^d - 1)A$ are the 'secants of dth degree' of A, i.e. the vectors $a_{i+d} - a_i$. If $d | n$, $\mathrm{Im}(\zeta^d - 1) = \mathfrak{A}_n^d$, and $\mathrm{Ker}(\zeta^d - 1) = \mathrm{Fix}\ \zeta^d = \mathfrak{A}_{d,\bar{d}}$.

2
OMITTING AVERAGING PROJECTIONS

Again let $d \mid n$. The cyclic mapping $(a_1, a_2, ..., a_n) \rightarrow (b_1, b_2, ..., b_n)$ with

$$b_1 = (d/n)(a_1 + a_{d+1} + ... + a_{n-d+1}), ... \tag{3}$$

will be denoted by μ_d; μ_d is an isobaric cyclic mapping. $b_1, b_2, ..., b_d$ are the centres of gravity of the rows of the vertex scheme modulo d, and thus of the omitting sub-$\bar{d}$-gons of $(a_1, a_2, ..., a_n)$. Hence $b_1 = b_{d+1}, ...$, and $\mu_d(a_1, a_2, ..., a_n)$ *is a d-gon counted $\bar{d}$ times.*

THEOREM 1 μ_d *is a projection*; $\mathrm{Im}\ \mu_d = \mathfrak{A}_{d,\bar{d}}$; $\mathrm{Ker}\ \mu_d = \mathfrak{A}_n^d$.
                                                                                       σ

PROOF   As was noted previously, $\mathrm{Im}\ \mu_d \subseteq \mathfrak{A}_{d,\bar{d}}$. Moreover it is obvious that $\mathfrak{A}_{d,\bar{d}} \subseteq \mathrm{Fix}\ \mu_d$. (If $a_1 = a_{d+1}, ...$, then from (3) $b_1 = a_1, ...$) Then, since $\mathrm{Im}\ \mu_d \subseteq \mathfrak{A}_{d,\bar{d}} \subseteq \mathrm{Fix}\ \mu_d$, the first two statements of the theorem hold. The third statement is simply the definition of the mapping μ_d and of the class $\mathfrak{A}_n^d$:

$$\mu_d(a_1, a_2, ..., a_n) = \sigma(a_1, a_2, ..., a_n)$$

is simply another way of writing the cyclic system of equations (2) which defines the class $\mathfrak{A}_n^d$.

We see also that $\mu_d = (d/n)(1 + \zeta^d + \zeta^{2d} + \ldots + \zeta^{n-d})$. Special cases are $\mu_1 = \sigma$ and $\mu_n = 1$.

The set of the $\tau(n)$ cyclic projections μ_d is closed under multiplication: if (r, s) denotes the greatest common divisor (g.c.d.) of r and s, then we have the following rule:

RULE $\quad \mu_r \mu_s = \mu_{(r,s)}$ *for r, s* $\mid n$.

The proof is left as an exercise.

According to §3.3 the following theorem for the cyclic projection $1 - \mu_d + \sigma$ (the σ-complement of μ_d and also isobaric) follows from theorem 1:

THEOREM 2 $\quad \mathrm{Im}(1 - \mu_d + \sigma) = \mathfrak{A}_n^d; \ \underset{\sigma}{\mathrm{Ker}} \ (1 - \mu_d + \sigma) = \mathfrak{A}_{d,\bar{a}}.$

Exercise

Each of the eight free cyclic octagonal classes of §1.8 is the image of $\mathfrak{A}_8$ under one of these cyclic projections: $\mu_8 = 1, \mu_4, \mu_2, \mu_1 = \sigma, \mu_8 - \mu_4 + \mu_2, \mu_8 - \mu_4 + \mu_1, \mu_8 - \mu_2 + \mu_1,$ $\mu_4 - \mu_2 + \mu_1.$

3
THE COMPLEMENTARY PROJECTIONS

As in §2.4, we say that two cyclic projections are *complementary* when they can be derived from each other by the involutory operator $1 -$. Each cyclic projection φ has exactly one complementary cyclic projection, namely $1 - \varphi$.

Let us form the complements of the cyclic projections introduced in the last section and determine the images and kernels of them all together: $\mu_d, \ 1 - \mu_d + \sigma, \ 1 - \mu_d, \ \mu_d - \sigma$. For this purpose consider, in addition to the free cyclic classes $\mathfrak{A}_n^d, \mathfrak{A}_{d,\bar{a}}$, the associated zero-point classes $\mathfrak{A}_{\circ n}^d, \mathfrak{A}_{d,\bar{a}}$ (see §3.2), and verify the statements of the following scheme:

$1 - \mu_d + \sigma$	$1 - \mu_d$	Im	$\mathfrak{A}_n^d$	$\mathfrak{A}_{\circ n}^d$	Ker	$\mu_d - \sigma$	μ_d
μ_d	$\mu_d - \sigma$	$\longrightarrow$	$\mathfrak{A}_{d,\bar{a}}$	$\mathfrak{A}_{d,\bar{a}}$	$\longleftarrow$	$1 - \mu_d$	$1 - \mu_d + \sigma$

PROOF OF THESE STATEMENTS We know $\mathrm{Im}(1 - \mu_d + \sigma)$ and $\mathrm{Im}\ \mu_d$ from theorems 1 and 2. By the second rule of §3.2, $\mathrm{Im}(1 - \mu_d)$ and $\mathrm{Im}(\mu_d - \sigma)$ are the associated zero-point classes. The statements concerning the kernels are easily obtained by remembering that $\mathrm{Ker}\ \varphi = \mathrm{Im}(1 - \varphi)$ for any cyclic projection φ.

If we now apply theorem 9′ of chapter 2 to μ_d and $1-\mu_d+\sigma$, we obtain from the above scheme

THEOREM 3 $\mathfrak{A}_n = \mathfrak{A}_{d,\bar{d}} \oplus \overset{\circ}{\mathfrak{A}}{}^d_n;\ \mathfrak{A}_n = \mathfrak{A}^d_n \oplus \overset{\circ}{\mathfrak{A}}_{d,\bar{d}}.$

In V, as the first statement of theorem 3 says, *every n-gon is uniquely representable as the sum of a d-gon counted d̄ times and an n-gon whose omitting sub-d-gons*

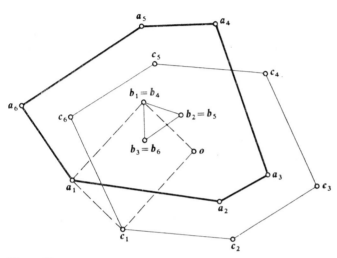

Figure 48

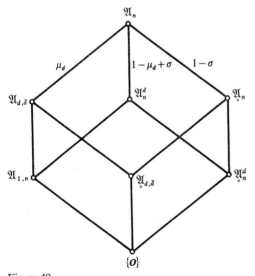

Figure 49

all have the same centre of gravity o. The first summand is the image of A under μ_d. If we form the centres of gravity of the omitting sub-$\tilde{d}$-gons of A, and from this d-tuple of centres of gravity we form a periodic n-gon B by $\tilde{d}$ repetitions, then $A = B+(A-B)$ is the required representation. For example, if $n = 2m$ and $d = m$, B is the doubly counted m-gon whose vertices are the mid-points of the 'diagonals' (pairs of opposite vertices) of A, and $A-B$ is a $2m$-parallelogram with centre of gravity o, i.e. a $2m$-gon with o as the common mid-point of its 'diagonals' (figure 48). In this case the maximum dimensions of B and $A-B$ are $m-1$ and m.

Figure 49 shows the inclusion diagram of the four cyclic classes of theorem 3 and the four basic classes (see §1.4). In this diagram, if two classes are joined by a segment, then one of the three projections μ_d, $1-\mu_d+\sigma$, $1-\sigma$ projects the upper class onto the lower.

Exercises

1 Every quadrangle is uniquely representable as the sum of a doubly counted digon and a parallelogram with centre of gravity o. Every hexagon is uniquely representable as the sum of a doubly counted triangle and a 6-parallelogram with centre of gravity o. Every hexagon is uniquely representable as the sum of a triply counted digon and an ASO hexagon with centre of gravity o.

2 If φ is an isobaric cyclic projection, then $\varphi-\sigma$ and $1-\varphi$ are non-isobaric projections with sum $1-\sigma$ and product 0. We have $\mathfrak{A}_n = \mathrm{Im}(\varphi-\sigma) \oplus \mathrm{Im}(1-\varphi) = \mathrm{Im}(\varphi-\sigma)$ $\oplus \mathrm{Ker}\ \varphi$ and, for $\varphi = \mu_d$, $\underset{\sigma}{\mathfrak{A}}_n = \mathfrak{A}_{d,\tilde{d}} \oplus \underset{\sigma}{\mathfrak{A}}_n^d$.

4

CONSECUTIVE AVERAGING PROJECTIONS

It seems sensible to consider consecutive averaging mappings as generalizations of the mappings κ_2 and κ_3 which were introduced earlier. Looked at from a systematic point of view, they are less important than the omitting averaging mappings. We denote by κ_d the cyclic mapping $(a_1, a_2, ..., a_n) \rightarrow (b_1, b_2, ..., b_n)$ with

$$b_1 = (1/d)(a_1+a_2+...+a_d), ...; \qquad (4)$$

κ_d is an isobaric cyclic mapping.

For $n = 4$, $\mathrm{Im}\ \kappa_2 = \mathrm{Ker}\ \underset{\sigma}{\mu_2}$ is the class $\mathfrak{A}_4^2$ of parallelograms, and $\mathrm{Im}\ \mu_2 =$ $\mathrm{Ker}\ \underset{\sigma}{\kappa_2}$ is the class $\mathfrak{A}_{2,2}$ of doubly counted digons. In general, for $d\mid n$

THEOREM 4 $\mathrm{Im}\ \kappa_d = \mathfrak{A}_n^d$; $\mathrm{Ker}\ \kappa_d = \underset{\sigma}{\mathfrak{A}}_{d,\tilde{d}}$.

To prove the first statement write the vertex scheme modulo d for $(b_1, b_2, ..., b_n)$.

The sum of the elements of any row is $(1/d)\sum a_i$. Thus the rows are isobaric and Im $\kappa_d \subseteq \mathfrak{A}_n^d$.

To prove the second statement note that $\kappa_d(a_1, a_2, ..., a_n) = \sigma(a_1, a_2, ..., a_n)$ says, when written as a cyclic system of equations,

$$(1/d)(a_1 + a_2 + ... + a_d) = (1/n)\sum a_i, \quad ..., \tag{5}$$

and this equation is equivalent to $a_1 = a_{d+1}, ...$ For from the first two equations of (5) we get $a_1 = a_{d+1}$, and, if $a_1 = a_{d+1}, ...$, then (5) is obviously fulfilled.

Note that $\kappa_d = (1/d)(1 + \zeta + \zeta^2 + ... + \zeta^{d-1})$ and $\kappa_d \mu_d = \sigma$.

To complete the proof of theorem 4, it is sufficient to show that $\mathfrak{A}_n^d \subseteq \mathrm{Im}\ \kappa_d$, i.e. that there is a pre-image under κ_d for every n-gon which is d times isobarically split. We shall prove even more:

(*) *For every n-gon in $\mathfrak{A}_n^d$, there exists exactly one κ_d-pre-image in $\mathfrak{A}_n^d$.*

PROOF   In $K[\zeta]$ there exists an element λ_d with

$$1 = \kappa_d \lambda_d + \mu_d \quad \text{(from (6), } \kappa_d \lambda_d \mu_d = 0). \tag{6}$$

One such element is

$$-(1 - \zeta) \cdot (d/n)(d\zeta^d + 2d\zeta^{2d} + ... + (n-d)\zeta^{n-d}).$$

This can be checked by using $\kappa_d(1 - \zeta) = (1/d)(1 - \zeta^d)$, $\zeta^n = 1$, and $\mu_d = (d/n)(1 + \zeta^d + ... + \zeta^{n-d})$. Consider now the cyclic mapping $\bar{\kappa}_d := \lambda_d + \sigma$. Because $\kappa_d \sigma = \sigma$ (chapter 3, (1)), $\kappa_d \bar{\kappa}_d = \kappa_d \lambda_d + \sigma$, and hence, from (6),

$$\kappa_d \bar{\kappa}_d = 1 - \mu_d + \sigma. \tag{7}$$

By theorem 2, $\mathfrak{A}_n^d = \mathrm{Fix}(1 - \mu_d + \sigma)$. Therefore

$$\mathfrak{A}_n^d = \mathrm{Fix}\ \kappa_d \bar{\kappa}_d. \tag{8}$$

Thus for any n-gon B in $\mathfrak{A}_n^d$ we have constructed a κ_d-pre-image, namely the n-gon $\bar{\kappa}_d B$; by theorem 5 of chapter 2 it lies in $\mathfrak{A}_n^d$. From (8) we see also that κ_d acts one-one on $\mathfrak{A}_n^d$: for if $A_1, A_2 \in \mathfrak{A}_n^d$ and if $\kappa_d A_1 = \kappa_d A_2$, then $\bar{\kappa}_d \kappa_d A_1 = \bar{\kappa}_d \kappa_d A_2$, and, because of the commutativity of cyclic mappings, (8) shows us that $A_1 = A_2$. Thus (*) and theorem 4 are completely proved.

When restricted to the class of n-gons which are d times isobarically split, the cyclic mapping $\bar{\kappa}_d$ constructed in this proof is the inverse of the restriction of κ_d.

COROLLARY κ_d *is a quasi-projection which has the same image and kernel as the cyclic projection* $1 - \mu_d + \sigma$.

This follows from theorem 4, (*), theorem 2, and the first rule in §3.2.

Exercises

1 $\bar{\kappa}_d$ is an invertible element in $K[\zeta]$. Determine its inverse.

2 κ_m, restricted to the class of $2m$-parallelograms, is one–one, and $\frac{1}{2}m(1-\zeta)+\sigma$ is the inverse of κ_m. For example, for $n = 4$, $1-\zeta+\sigma$ maps every parallelogram onto its circumscribed parallelogram (cf. §2.6, exercise 3 of §2.7).

3 If $n = d\bar{d}$ is a factorization of n into relatively prime factors, then

$$\mathrm{Ker}(\zeta^d - 1)(\zeta^{\bar{d}} - 1) = \underset{\sigma}{\mathrm{Ker}}\ \kappa_d\kappa_{\bar{d}} = \mathrm{Im}\ \mu_d + \mathrm{Im}\ \mu_{\bar{d}} = \mathfrak{A}_{d.\bar{d}} + \mathfrak{A}_{\bar{d}.d}$$

is the class of $(d, \bar{d})$-prisms (cf. §1.8, exercise 5).

PART II
The main theorem

5
Idempotent elements and Boolean algebras

The concept of a Boolean algebra and other lattice theory concepts which we shall use below are explained in appendix 1, which the reader is advised to read as a supplement to the text.

1

IDEMPOTENT ELEMENTS OF A RING

Let R, $+$, $\cdot$ be a commutative ring with elements 0, a, b, ... An element a is said to be *idempotent* if $aa = a$. We shall denote the set of idempotent elements of R by $E(R)$. $E(R)$ contains the zero element, and, if it contains a and b, it contains ab, but not always $a+b$.

For arbitrary elements of R we introduce a *circle composition* (also called a 'star product' by van der Waerden[1])

$$a \circ b := a+b-ab. \tag{1}$$

This circle composition is commutative and associative in R; 0 is the neutral element. If a, b are elements of R with $ab = 0$, then $a \circ b = a+b$. Such elements a, b with $ab = 0$ are said to be *orthogonal* to each other. If a, $b \in E(R)$, then $a \circ b \in E(R)$; in particular, note that $a \circ a = a$.

$E(R)$, $\circ$, $\cdot$ is a distributive lattice. The absorption and distributive laws can be easily verified. In doing so, we interpret $\circ$ as the supremum and $\cdot$ as the infimum, and define $a \leq b$ by $ab = a$, or equivalently by $a \circ b = b$. Then 0 is the least element of the lattice $E(R)$.

Let us now assume that R has a unity 1. Then $a \to 1-a$ is an involutory mapping of R into R, and so a one–one mapping of R onto R.

1 is idempotent and the greatest element of $E(R)$. If $a \in E(R)$, so does $1-a$ (and conversely). Moreover,

$$1 = a \circ (1-a), \quad a(1-a) = 0 \qquad \text{for all } a \in E(R). \tag{2}$$

Thus a and $1-a$ are complements in the lattice $E(R)$.

1 B.L. van der Waerden, *Modern Algebra* II (New York, 1950); trans. from the 2nd rev. German ed. *Algebra* II (Berlin, etc., 1949); N. Jacobson, *Structure of Rings* (Providence, 1964), uses circle composition.

Thus $E(R)$, $\circ$, $\cdot$ is a distributive, complemented lattice and so a Boolean algebra.

THEOREM 1   *In any commutative ring with* 1, *the idempotent elements form a Boolean algebra with respect to* $\circ$, $\cdot$.

We shall sometimes call $\circ$ the 'Boolean sum'.
The operator $1-$ maps R, $\circ$, $\cdot$ onto R, $\cdot$, $\circ$, for

$$1-(a \circ b) = (1-a)(1-b), \quad 1-ab = (1-a) \circ (1-b) \qquad \text{for all } a, b \in R. \qquad (3)$$

If we set $1-a = a'$, then $a'' = a$, and equations (3) become the de Morgan laws:

$$(a \circ b)' = a'b', \quad (ab)' = a' \circ b' \qquad \text{for all } a, b \in R.$$

EXAMPLES OF THEOREM 1
  1. In an integral domain R with 1 (and more particularly in a field), 0, 1 are the only idempotent elements. For $a^2 = a$ implies that $a(1-a) = 0$ and, since R has no divisors of zero, $a = 0$ or $1-a = 0$.
  2. The idempotents of the residue class ring (or 'quotient ring') $Z/(30)$ are 0, 1, 6, 10, 15, 16, 21, 25. Figure 50 exhibits the Boolean algebra $E(Z/(30))$.
  3. Let $K_1, K_2, ..., K_k$ be fields. Let $R = \Sigma \oplus K_i$, so that R is the set of k-tuples

$$(a_1, a_2, ..., a_k) \quad \text{with } a_i \in K_i \qquad (4)$$

which are added and multiplied by components. Then $E(R)$ consists of the 2^k elements (4) with $a_i \in \{0, 1\}$. The k elements

$$(1, 0, ..., 0), \quad (0, 1, 0, ..., 0), \quad ..., \quad (0, ..., 0, 1) \qquad (5)$$

are the atoms of $E(R)$. The product of any two distinct elements is $(0, 0, ..., 0)$, the zero element of R; they are 'orthogonal' to each other. The Boolean sum of

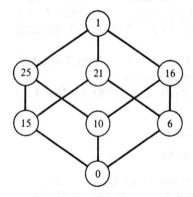

Figure 50

all the elements (5) is the usual sum and is equal to $(1, 1, ..., 1)$, the unit element of R. (Example 2 may be thought of as a special case of this.)

Exercises

1 Let R be a commutative ring with 1. If $a, b \in E(R)$, then

$$a \oplus b : = a+b-2ab = a \circ b - ab = ab' \circ a'b$$

is also a member of $E(R)$. $E(R)$, $\oplus$, $\cdot$ is a Boolean ring with 1, that is a ring with 1 in which every element is idempotent.

2 Let R be a commutative ring with 1 in which $\frac{1}{2}$ exists, and let $J(R)$ be the set of involutory elements in R, i.e. elements whose squares are 1. $J(R)$ is a subgroup of the group of units of R. The mapping (*) $e \to 2e-1$ is a one–one mapping of $E(R)$ onto $J(R)$. If, using this mapping, we transfer the lattice operations, $\circ$, $\cdot$ given on $E(R)$ onto $J(R)$, then $J(R)$ is a Boolean algebra with respect to

$$a \sqcup b : = \tfrac{1}{2}(1+a+b-ab), \qquad a \sqcap b : = \tfrac{1}{2}(-1+a+b+ab).$$

1 is the greatest element, -1 the least, and a, $-a$ are complementary. $\sqcup$, $\sqcap$ are the supremum and infimum with respect to the relation $a \leq b$ defined on $J(R)$ by $1+a = (1+a)b$.

If $e, f \in E(R)$, and if e is mapped onto a, f onto b by (*), then $ef \circ e'f'$ is mapped onto ab ($e' = 1-e$, $f' = 1-f$). ef, $e'f'$ are orthogonal.

3 If R is the direct sum of k fields K_i with characteristic $\neq 2$, then $J(R)$ consists of the 2^k elements (4) with $a_i \in \{1, -1\}$.

2
FINITELY GENERATED BOOLEAN ALGEBRAS

The examples given above are also examples of the following theorem which starts with a representation of 1 as a sum of mutually orthogonal elements.

THEOREM 2   *In a commutative ring R with 1, let $e_1, e_2, ..., e_k$ be elements $\neq 0$ with*

$$1 = e_1+e_2+...+e_k \tag{6}$$

and

$$e_i e_j = 0 \quad for \; i \neq j. \tag{7}$$

Then the partial sums of $e_1+e_2+...+e_k$ form, with respect to $\circ$, $\cdot$, a Boolean algebra whose atoms are the e_i and which has 2^k elements.

By the *partial sums* of

$$e_1+e_2+...+e_k \tag{8}$$

we mean (8) itself and all expressions obtained from it by omitting terms in all possible ways. Formally there are 2^k such partial sums, including the empty sum which is set equal to 0. The proof of theorem 2 is simply a series of verifications of properties. When we mention 'partial sums' we shall mean partial sums of (8).

1. Multiplying (6) by e_i we obtain $e_i = e_i^2$ (using (7)). Thus all the e_i's are idempotent.

2. Since the e_i's are mutually orthogonal, (8), and so every partial sum, can be written as a Boolean sum (with $\circ$ instead of $+$). In particular, all the partial sums are idempotent.

3. The product of two partial sums is a partial sum; it consists of the e_i's which occur in both partial sums (the 'intersection').

4. The Boolean sum of two partial sums is a partial sum; it consists of the e_i's which appear in at least one of the partial sums (the 'union'; for example,

$$(e_1 + e_2) \circ (e_2 + e_3) = e_1 \circ e_2 \circ e_2 \circ e_3 = e_1 \circ e_2 \circ e_3 = e_1 + e_2 + e_3).$$

5. The complement of a partial sum is the sum of the e_i's not appearing in the given partial sum.

From 1 and 2, the partial sums are elements of $E(R)$; by 3, 4, 5 they form a Boolean subalgebra of $E(R)$, $\circ$, $\cdot$.

6. Two partial sums are equal only if they consist of the same e_i's (e.g. from $e_1 + e_2 + e_3 = e_2 + e_3 + e_4$ by multiplication by e_1 we obtain $e_1 = 0$; but by assumption $e_1 \neq 0$). In particular, $e_i \neq e_j$ if $i \neq j$. The subalgebra has 2^k elements.

7. The product of e_i and an arbitrary partial sum is e_i or 0, according to whether e_i appears in that partial sum or not (see 3). From this it follows immediately that e_i is either an atom of the subalgebra or equal to 0, but since we assumed that $e_i \neq 0$, e_i is an atom. Finally it is clear that partial sums with more than one e_i are not atoms. Theorem 2 is thus proved.

COROLLORY 1 *Without the assumption that* $e_1, e_2, ..., e_k \neq 0$, *the partial sums* $e_1 + e_2 + ... + e_k$ *still form a Boolean subalgebra of* $E(R)$, $\circ$, $\cdot$. *The atoms are now the* $e_i \neq 0$. *If l is the number of non-zero* e_i's, *then the subalgebra has* 2^l *elements.*

COROLLARY 2 *For elements* $e_1, e_2, ..., e_k \in R$, (6) *and* (7) *are equivalent to*

$$1 = e_1 \circ e_2 \circ ... \circ e_k, \tag{6'}$$

$$e_i(e_1 \circ e_2 \circ ... \circ e_{i-1} \circ e_{i+1} \circ ... \circ e_k) = 0. \tag{7'}$$

The statement that $e_1, e_2, ..., e_k$ are mutually orthogonal elements with sum 1 is thus equivalent to saying that each element $e_1, e_2, ..., e_k$ is complementary to the Boolean sum of the others, i.e. that 1 is the 'direct' Boolean sum of $e_1, e_2, ..., e_k$.

PROOF (6') and (7') follow from (6) and (7), as the proof of theorem 2 shows (see

1, 2, 3). Conversely, let elements $e_1, e_2, ..., e_k \in R$ be given for which (6'), (7') hold. With the abbreviation $e_1 \circ ... \circ e_{i-1} \circ e_{i+1} \circ ... \circ e_k = e_i{}^*$, we have

$$1 = e_i \circ e_i{}^* \quad \text{and} \quad e_i e_i{}^* = 0.$$

From this equation we get $1 = e_i + e_i{}^*$. Multiplication by e_i gives $e_i = e_i{}^2$ so that all the e_i's are idempotent and their Boolean sums are also idempotent. If $i \neq j$, $e_i{}^* e_j = e_j$ (since for any $e \in E(R)$, $(e \circ e_j)e_j = e_j$ by the absorption law). Multiplying by e_i yields $e_i e_j = 0$, and so (7) holds. (6) follows from (7) and (6').

   Now let $e_1, e_2, ..., e_k$ be arbitrary elements from $E(R)$ ((6) and (7) need not hold). Form the so-called *Boolean minimal polynomials in* $e_1, e_2, ..., e_k$:

$$e_1 e_2 ... e_k \tag{9}$$

is one, and the others are obtained from it by replacing factors e_i by e_i' in all possible ways $(e_i' : = 1 - e_i)$. Formally, there are 2^k such minimal polynomials: $M_1, M_2, ..., M_{2^k}$. We have

$$1 = M_1 + M_2 + ... + M_{2^k} \tag{10}$$

and

$$M_i M_j = 0 \quad \text{for } i \neq j. \tag{11}$$

((10) is proved by induction on k.) *The sums of the minimal polynomials, i.e. the partial sums of*

$$M_1 + M_2 + ... + M_{2^k}, \tag{12}$$

form a Boolean subalgebra of $E(R)$, $\circ$, $\cdot$ by corollary 1. This subalgebra contains the elements $e_1, e_2, ..., e_k$. (E.g. write $e_1 = e_1 \cdot 1$ and then replace the factor 1 by the sum of all minimal polynomials in $e_2, ..., e_k$.) Obviously this is the smallest Boolean subalgebra of $E(R)$ which contains the given elements, and so is the Boolean subalgebra generated by $e_1, e_2, ..., e_k$. The minimal polynomials which are different from 0 are the atoms of the subalgebra. If l is the number of them, the subalgebra has 2^l elements. Thus $l \leq 2^k$. If $l = 2^k$, then we say that $e_1, e_2, ..., e_k$ generate the subalgebra *freely*.

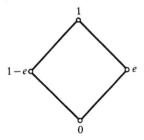

Figure 51

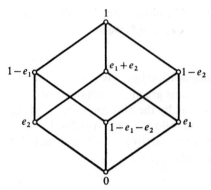

Figure 52

EXAMPLES

1. Let $e \in E(R)$ and $e \neq 0, 1$. The minimal polynomials in e are e, e'; they are not equal to zero. The only other sums of minimal polynomials are 0 and $e + e' = 1$. The Boolean algebra generated by e consists of $0, 1, e, e'$. (See figure 51.)

2. Let $e_1, e_2 \in E(R)$, and let $e_1, e_2 \neq 0$, $e_1 e_2 = 0$, $e_1 + e_2 \neq 1$. The minimal polynomials in e_1, e_2 are $e_1 e_2 = 0, e_1 e_2' = e_1, e_1' e_2 = e_2, e_1' e_2' = 1 - e_1 - e_2$. The last three are $\neq 0$. Adding these minimal polynomials we also obtain $0, e_1 + e_2$, $e_1', e_2', 1$. The eight elements of the Boolean algebra generated by e_1 and e_2 are thus accounted for. (See figure 52.)

3

IDEMPOTENT ENDOMORPHISMS OF AN ABELIAN GROUP: Im-TRANSFER

In subsequent development of our theory certain theorems which we shall call transfer theorems will play a special role. By a *transfer* we mean an isomorphic or anti-isomorphic mapping of a lattice L_1 onto a sublattice of a lattice L_2, i.e. a one–one mapping of L_1 into L_2 which either preserves or interchanges the compositions supremum and infimum. In the decisive theorems it will be a matter of injecting a 'small' lattice with special properties, say a finite Boolean algebra, into a 'large' lattice which in general will not be distributive.

Now let $\mathfrak{A}$, $+$ be an Abelian group, End($\mathfrak{A}$), $+$, $\cdot$ the ring of endomorphisms of $\mathfrak{A}$, and $L(\mathfrak{A})$, $+$, $\cap$ the lattice of subgroups of $\mathfrak{A}$.

Let us start from §2.4, theorem 8, which can be interpreted as a transfer theorem: if φ is an idempotent endomorphism of $\mathfrak{A}$, then

$$\{0, 1, \varphi, 1 - \varphi\}, \circ, \cdot \tag{13}$$

is a Boolean algebra of endomorphisms of $\mathfrak{A}$. The theorem mentioned above says: Im is an isomorphism of the Boolean algebra (13) onto the sublattice

$$\{\{o\}, \mathfrak{A}, \text{Im } \varphi, \text{Im}(1 - \varphi)\}, +, \cap \tag{14}$$

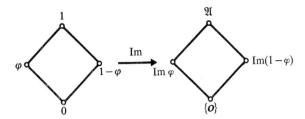

Figure 53

of the lattice $L(\mathfrak{A})$; (14) is, therefore, also a Boolean algebra (see figure 53). Dually Ker is an anti-isomorphism of (13) onto (14).

This statement can be generalized. Consider an arbitrary Boolean algebra $E, \circ, \cdot$ of endomorphisms of $\mathfrak{A}$. Note that when we speak of a Boolean algebra E of endomorphisms of $\mathfrak{A}$, this includes the statement that all elements of E are idempotent elements of End($\mathfrak{A}$) and commute in pairs. For such elements, $\varphi \leq \psi$ is defined by $\varphi\psi = \varphi$ or equivalently $\varphi \circ \psi = \psi$.

THEOREM 3 (Im-transfer)   *If $E, \circ, \cdot$ is a Boolean algebra of endomorphisms of an abelian group $\mathfrak{A}$, then* Im *(the mapping $\varphi \to$ Im φ) is an isomorphism of $E, \circ, \cdot$ onto a sublattice of the subgroup lattice $L(\mathfrak{A}), +, \cap$.*

Naturally this sublattice is also a Boolean algebra and will be denoted by $L_E(\mathfrak{A})$.

By chapter 2, theorem 7, E is mapped one–one by Im. To complete the proof of theorem 3 we need only show that for commuting idempotent endomorphisms φ, ψ

$$\text{Im}(\varphi \circ \psi) = \text{Im } \varphi + \text{Im } \psi, \tag{15}$$

$$\text{Im } \varphi\psi = \text{Im } \varphi \cap \text{Im } \psi. \tag{16}$$

PROOF OF (15) AND (16)   If $\varphi, \psi \in$ End($\mathfrak{A}$) and $\varphi\psi = \psi\varphi$, then

$$\text{Fix } \varphi + \text{Fix } \psi \subseteq \text{Fix}(\varphi \circ \psi) \subseteq \text{Im}(\varphi \circ \psi) \subseteq \text{Im } \varphi + \text{Im } \psi, \tag{15'}$$

$$\text{Fix } \varphi \cap \text{Im } \psi \subseteq \text{Im } \varphi\psi \subseteq \text{Im } \varphi \cap \text{Im } \psi. \tag{16'}$$

For the first inclusion of (15'), we see that if $\varphi a = a$ then $(\varphi \circ \psi)a = (\varphi + \psi - \psi\varphi)a = a$. Therefore Fix $\varphi \subseteq$ Fix($\varphi \circ \psi$). The second inclusion holds generally (see §2.4). For the third inclusion, $(\varphi \circ \psi)a = (\varphi + \psi - \psi\varphi)a = \varphi a + \psi(a - \varphi a)$.

For the first inclusion of (16'), if an element $\psi a \in$ Fix φ, then $\varphi\psi a = \psi a$, and thus $\psi a \in$ Im $\varphi\psi$. The second inclusion follows from $\varphi\psi a = \varphi(\psi a) = \psi(\varphi a)$.

If, moreover, φ and ψ are idempotent, then Fix $\varphi =$ Im φ, Fix $\psi =$ Im ψ, and the equations (15), (16) follow from (15'), (16'). ((16') holds for commuting φ, ψ if at least one is idempotent.)

If the endomorphisms φ, ψ commute, the following rules also hold:

LIBRARY
OF
MOUNT ST. MARY'S
COLLEGE
EMMITSBURG, MARYLAND

$$\text{Ker } \varphi\psi = \text{Ker } \varphi + \text{Ker } \psi, \tag{17}$$

$$\text{Ker}(\varphi \circ \psi) = \text{Ker } \varphi \cap \text{Ker } \psi. \tag{18}$$

These are proved from (15), (16) with the help of the equation $\text{Ker } \varphi = \text{Im}(1-\varphi)$ (valid for every idempotent endomorphism φ) and the de Morgan laws. The following three statements are equivalent to one another:

(i) $\varphi \le \psi$,
(ii) $\text{Im } \varphi \subseteq \text{Im } \psi$,
(iii) $\text{Ker } \varphi \supseteq \text{Ker } \psi$.

Theorem 3 is of interest for us when $0, 1 \in E$. For then, if E contains an element φ, it also contains its complement $1 - \varphi$. Then $\{o\}$, $\mathfrak{A}$ lie in $L_E(\mathfrak{A})$; under Im-transfer complementary elements of E give complementary subgroups of $\mathfrak{A}$, just as in the special case of theorem 3 mentioned above; the combined mapping $\varphi \to 1-\varphi \to \text{Im}(1-\varphi) = \text{Ker } \varphi$, and thus Ker, is an anti-isomorphism of E onto $L_E(\mathfrak{A})$.

COROLLARY   *If $\varphi_1, \varphi_2, \ldots, \varphi_k$ are endomorphisms of $\mathfrak{A}$ with*

$$1 = \varphi_1 + \varphi_2 + \ldots + \varphi_k \tag{19}$$

and

$$\varphi_i \varphi_j = 0 \quad \text{for } i \ne j, \tag{20}$$

then $\mathfrak{A} = \sum \oplus \text{Im } \varphi_i$.

PROOF   By (20), $\varphi_1, \varphi_2, \ldots, \varphi_k$ commute in pairs. Therefore in the subring of $\text{End}(\mathfrak{A})$ generated by them, we may use theorem 2 and its corollaries: the partial sums of $\varphi_1 + \varphi_2 + \ldots + \varphi_k$ form a Boolean algebra whose atoms are the $\varphi_i \ne 0$. If we replace (19), (20) by the equivalent equations in the second corollary, then Im-transfer (theorem 3) yields the assertion.

Exercises

1 Determine the Boolean algebra of the idempotent endomorphisms for the cyclic groups of orders 6, 30, and 105.
2 If φ is an idempotent endomorphism of a vector space over a field of characteristic $\ne 2$, then $1 + \varphi$ is an automorphism. What is its inverse?

4

THE BOOLEAN ALGEBRA OF CYCLIC PROJECTIONS

The main point to be learned from the theorems of this chapter for the theory of n-gons is the following: we can apply theorem 1 to $K[\zeta]$, the algebra of cyclic mappings, and consider

E(K[ζ]), ∘, · the Boolean algebra of cyclic projections.

By theorem 3 we can isomorphically transfer this Boolean algebra into the lattice of subspaces of the vector space of n-gons, and obtain a Boolean algebra of cyclic classes (chapter 2, theorem 9). In chapter 6 we shall see that the Boolean algebra of cyclic classes thus obtained contains all cyclic classes.

From §2.4 we know that, of two complementary projections of $E(K[ζ])$, one is always isobaric while the other has coefficient sum zero. Of the two complementary cyclic classes arising by Im-transfer, one is a free class and the other a zero-point class (see chapter 2, theorem 9′).

σ is an atom of E(K[ζ]). (For a cyclic projection φ we always have $s(\varphi) \in \{0, 1\}$. Thus, by the lemma of §3.1, $\varphi\sigma = 0$ or $\varphi\sigma = \sigma$. If $\varphi \leq \sigma$, i.e. $\varphi\sigma = \varphi$, then $\varphi = 0$ or $\varphi = \sigma$.) Since σ is the least isobaric cyclic projection (chapter 3, (1)), other atoms of $E(K[ζ])$ must necessarily have coefficient sum 0. In the Boolean algebra of cyclic classes arising by Im-transfer from $E(K[ζ])$ the class Im $\sigma = \mathfrak{A}_{1,n}$ of trivial n-gons, the least free cyclic class, is an atom and the other atoms are zero-point classes.

The Boolean algebra $E(K[ζ])$ contains at least the $\tau(n)$ omitting averaging projections μ_d ($d\,|\,n$) and the Boolean subalgebra generated by them. (Note that $\mu_1 = \sigma, \mu_n = 1$.)

5
EXAMPLES OF Im-TRANSFER

The Boolean subalgebra of $E(K[ζ])$ generated by σ consists only of 0, 1, σ, $1-\sigma$. Using Im-transfer we obtain the Boolean algebra of the four cyclic basic classes (cf. §1.4 and see figure 54).

Next we shall clarify the assertions of §4.3 from the present point of view, and for a single non-trivial divisor d of n form *the Boolean subalgebra $E(\mu_d, σ)$ of $E(K[ζ])$ generated by the cyclic projections* μ_d, σ. The Boolean minimal polynomials in μ_d, σ are

$$\mu_d\sigma = \sigma, \quad (1-\mu_d)\,\sigma = 0, \quad \mu_d(1-\sigma) = \mu_d - \sigma, \quad (1-\mu_d)(1-\sigma) = 1-\mu_d. \quad (21)$$

We always have $\sigma \neq 0$, and, since $d \neq 1, n$, we also have $\mu_d - \sigma, 1 - \mu_d \neq 0$. The three minimal polynomials $\neq 0$ are the atoms of the Boolean algebra $E(\mu_d, \sigma)$. The

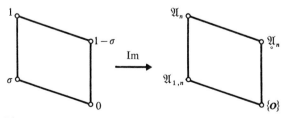

Figure 54

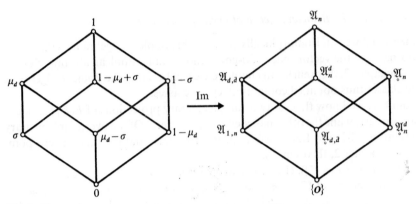

Figure 55

Boolean algebra consists of the $2^3 = 8$ sums of minimal polynomials $\neq 0$. They are 0, the atoms, the complements $1-\sigma$, $1-\mu_d+\sigma$, μ_d of the atoms, and 1. (This holds not only for μ_d but also for an arbitrary isobaric cyclic projection $\varphi \neq 1, \sigma$.)

Figure 55 shows the diagram of the Boolean algebra $E(\mu_d, \sigma)$. The diagram of the cyclic classes arising through Im-transfer is known from figure 49 (§4.3), and we now know that these eight cyclic classes form a Boolean algebra with respect to sum and intersection.

It is more interesting to search for the *Boolean subalgebra* E_μ of $E(K[\zeta])$ *generated by all cyclic projections* μ_d $(d \mid n)$ for a given n. For $n = 4$, or more generally $n = p^2$ (p a prime), this is of the type just mentioned above. Let us now consider in detail the case $n = 6$ and study the Boolean algebra E_μ generated by $\mu_1 = \sigma$, μ_2, μ_3, $\mu_6 = 1$. It is generated by μ_2, μ_3. The minimal polynomials in μ_2, μ_3 are

$$\mu_2\mu_3 = \sigma, \quad \mu_2(1-\mu_3) = \mu_2-\sigma, \quad (1-\mu_2)\mu_3 = \mu_3-\sigma, \quad (1-\mu_2)(1-\mu_3). \quad (22)$$

These are distinct from 0 and are thus the atoms of the subalgebra E_μ which consists of the $2^4 = 16$ sums of minimal polynomials. These 16 cyclic projections group themselves into eight which are isobaric and eight with coefficient sum zero. The eight isobaric ones are

$$1 \quad 1-\mu_2+\sigma \quad 1-\mu_3+\sigma \quad (1-\mu_2+\sigma)(1-\mu_3+\sigma)$$
$$\sigma \qquad \mu_2 \qquad \mu_3 \qquad \mu_2+\mu_3-\sigma \qquad\qquad (23)$$

(σ-complementary pairs are written in columns). The non-isobaric pairs are obtained from this by subtracting σ (or multiplying by $1-\sigma$; see chapter 3, (2)).

THEOREM 4   *Let* $n = 6$. *The Boolean algebra generated by the* $\tau(6) = 4$ *omitting averaging projections consists of* $2^4 = 16$ *cyclic projections. The images of* $\mathfrak{A}_6$ *under these projections are the 8 free cyclic hexagonal classes of §1.8 and the associated zero-point classes. These 16 cyclic hexagonal classes form a Boolean algebra in the lattice of subspaces of* $\mathfrak{A}_6$, *the vector space of hexagons.*

PROOF   We must prove the second statement; the last follows from theorem 3. The images of $\mathfrak{A}_6$ under the isobaric cyclic projections (23) are the eight cyclic hexagonal classes of §1.8: the images under the first three pairs are easily found using theorems 1 and 2 of chapter 4; moreover, by §4.1, $\mathrm{Im}(1 - \mu_2 + \sigma)(1 - \mu_3 + \sigma)$ $= \mathrm{Im}(1 - \mu_2 + \sigma) \cap \mathrm{Im}(1 - \mu_3 + \sigma) = \mathfrak{A}_6^2 \cap \mathfrak{A}_6^3$, the class $\mathfrak{R}_6$ of affinely regular hexagons, and $\mathrm{Im}(\mu_2 + \mu_3 - \sigma) = \mathrm{Im}(\mu_2 \circ \mu_3) = \mathrm{Im}\ \mu_2 + \mathrm{Im}\ \mu_3 = \mathfrak{A}_{2,3} + \mathfrak{A}_{3,2}$, the class of prisms. The images of $\mathfrak{A}_6$ under the eight non-isobaric cyclic projections of E_μ are the zero-point classes associated with these eight free cyclic classes (second rule from §3.2).

From the representation of 1 as sum of the atoms (22) of E_μ,

$$1 = \sigma + (\mu_2 - \sigma) + (\mu_3 - \sigma) + (1 - (\mu_2 + \mu_3 - \sigma)), \tag{24}$$

we get, by Im-transfer (corollary to theorem 3),

$$\mathfrak{A}_6 = \mathfrak{A}_{1,6} \oplus \mathfrak{A}_{2,3} \oplus \mathfrak{A}_{3,2} \oplus \mathfrak{R}_6. \tag{25}$$

On the right-hand side we have the atomic classes of the Boolean algebra of 16 cyclic hexagonal classes (cf. figure 46). (25) says:

THEOREM 5   *Every hexagon is uniquely representable as the sum of a trivial hexagon, a triply counted digon with centre of gravity o, a doubly counted triangle with centre of gravity o, and an affinely regular hexagon with centre of gravity o.*

Applying (24) to a hexagon A gives

$$A = \sigma A + (\mu_2 - \sigma)A + (\mu_3 - \sigma)A + (1 - (\mu_2 + \mu_3 - \sigma))A \tag{26}$$

as the decomposition of A into these four '*components.*'

σA, the first component of A, is the centre-of-gravity hexagon of A (i.e. the centre of gravity of A counted 6 times); $\mu_2 A$ is the triply counted digon formed by the centres of gravity of the two omitting triangles of A; $\mu_3 A$ is the doubly counted triangle formed by the mid-points of the diagonals of A (the mid-points of opposite vertices). If $\mu_2 A$ and $\mu_3 A$ are 'shifted to the origin' (i.e. σA is subtracted from them) we then obtain the second and third components of A (see figure 56).

The hexagon $(\mu_2 + \mu_3 - \sigma)A$ is constructed by forming the fourth vertex of the parallelogram determined by the ith vertex of $\mu_2 A$, the centre of gravity of A, and the ith vertex of $\mu_3 A$. As we know, a prism is obtained. This prism, associated with the hexagon A in a geometrically natural way, is called *the prism spanned by the centres of gravity of the subpolygons of* A. Let us denote it by A^*, and note that, for a prism A, $A^* = A$.

$(1 - (\mu_2 + \mu_3 - \sigma))A = A - A^*$, the fourth component of A, is an affinely regular hexagon with centre of gravity o. We formulate the geometrical relations which yield the construction of this '*affinely regular component of* A' as

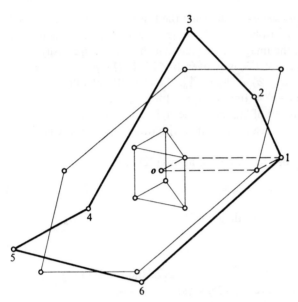

Figure 56 Decomposition of a 6-gon with centre of gravity o into a prism with centre of gravity o and an affinely regular 6-gon with centre of gravity o

THEOREM 6 *If A is an arbitrary hexagon and A^* is the prism spanned by the centres of gravity of the subpolygons of A, then $A - A^*$ is an affinely regular hexagon with centre of gravity o.*

Thus, if $A = (a_1, ..., a_6)$, $A^* = (a_1{}^*, ..., a_6{}^*)$, then the position vectors $a_1 - a_1{}^*, ..., a_6 - a_6{}^*$ give the vertices of an affinely regular hexagon. A is in general 5-dimensional, A^* at most 3-dimensional. The fact that $A - A^*$ is at most 2-dimensional seems worth mentioning.

Exercises

1 Perform the corresponding investigation for $n = 8$.
2 The Boolean algebra generated by all μ_d $(d \mid n)$ has $\tau(n)$ atoms.
3 Let $K = Q$, $V = Q^2$. Decompose the hexagon $((-5, 6), (7, 7), (10, 2), (0, -7), (-5, -5), (-7, -3))$ according to theorem 5.
4 Every quadrangle is uniquely representable as the sum of a trivial quadrangle, a doubly counted digon with centre of gravity o, and a parallelogram with centre of gravity o. If $K = Q$, $V = Q^2$, decompose the quadrangle $((0, 0), (7, 1), (6, 6), (3, 7))$ in this way.

6
The main theorem about cyclic classes

1
CONGRUENCES IN PRINCIPAL IDEAL DOMAINS

First let R be a commutative ring with 1. The units (invertible elements) of R form an abelian group U with respect to multiplication. Elements a, $b \in R$ are called *associated* (in symbols $a \sim b$) if there exists $u \in U$ such that $ua = b$. This association is an equivalence relation on R. The class of associates in which a lies is Ua. (Note that a can be a zero divisor, and then the equation $xa = b$ for $b = 0$ can have several solutions for the one b. Thus, if a and b are associated, there may still be additional equations $ca = b$ with $c \notin U$.) The principal ideal Ra generated by a will be denoted as usual by (a). If $a \sim b$, then $(a) = (b)$.

LEMMA *If e, e^* are idempotent elements of a commutative ring with 1, then the following statements are equivalent*:

(a) $e \sim e^*$,
(b) $(e) = (e^*)$,
(c) $e = e^*$.

A class of associates contains at most one idempotent element.

PROOF (a) implies (b), and (c) implies (a). We must prove that (b) implies (c).

Since e is idempotent, every element in (e) remains unchanged under multiplication by e. Since $e^* \in (e)$, $e^*e = e^*$. But e^* is also idempotent and $e \in (e^*)$, so that $ee^* = e$. Thus $e = ee^* = e^*e = e^*$ since R is commutative.

By a *principal ideal domain* we mean *an integral domain with 1 in which every ideal is a principal ideal.*[1] Examples are the ring of integers and the polynomial rings over fields. In a principal ideal domain every non-zero element which is not a unit may be represented as a product of prime elements, and this representation is unique up to the order of the factors and the replacement of factors by associates.

---

1 See, e.g., R. Kochendoerffer, *Einführung in die Algebra* (Berlin, 1955), §49; B.L. van der Waerden, *Modern Algebra* I (New York, 1950), §§17, 18; A.G. Kurosh, *Lectures on General Algebra* (New York, 1963), trans. from the original Russian text *Lekcii po Obshchei Algebre* (Moscow, 1963), §§15, 16.

D

Let R be a principal ideal domain and m an element $\in R$. Define

$$a \equiv b \quad \mathrm{mod}(m) \qquad \text{by } a - b \in (m).$$

This congruence modulo (m) is an equivalence relation on R which is compatible under addition and multiplication in R.

An element $e \in R$ is called *modulo* (m) *idempotent* if $e^2 \equiv e \ \mathrm{mod}(m)$. This congruence is equivalent to $e(1-e) \equiv 0 \ \mathrm{mod}(m)$; thus, if e is modulo (m) idempotent, $e' : = 1 - e$ is also modulo (m) idempotent. The product and Boolean sum of two modulo (m) idempotent elements are modulo (m) idempotent.

An element $m \in R$ is said to be *square-free* if it is a product of prime elements no two of which are associated, thus if it possesses a representation

$$m = p_1 p_2 \ldots p_k \text{ with no two } p_i\text{'s associated } (k \le 1). \tag{*}$$

For the case in which m is a square-free element, we shall now prove two theorems concerning the modulo (m) idempotent elements of a principal ideal domain R. We assume that the tools used here for congruences are familiar to the reader from elementary number theory when R is the ring of integers. For the proof of the main theorem about cyclic classes of n-gons we need the related case, where R is the ring of polynomials over a field. However it is clearer to phrase the statements for an arbitrary principal ideal domain.

Let m always be of the form (*) and $i \in \{1, 2, \ldots, k\}$.

(i) $a \equiv b \ mod(m)$ *if and only if $a \equiv b$ modulo each* (p_i).
(ii) e *is modulo* (m) *idempotent if and only if $e \equiv 0$ or 1 modulo each* (p_i).

PROOF The following expressions are equivalent: $e(1-e) \equiv 0 \ \mathrm{mod}(m)$; $e(1-e) \equiv 0$ modulo each (p_i); $e \equiv 0$ or $e \equiv 1$ modulo each (p_i).

(iii) 'Chinese construction': *For every p_i there exists an e_i in R such that*

$$e_i \equiv 1 \quad \mathrm{mod}(p_i), \tag{1}$$

$$e_i \equiv 0 \quad \mathrm{mod}(m/p_i). \tag{2}$$

From (2) we get

$$e_i \equiv 0 \quad \mathrm{mod}(p_j) \qquad \text{for } i \neq j. \tag{2'}$$

By (ii), e_i *is modulo* (m) *idempotent and by* (i) *it is uniquely determined modulo* (m). *Moreover*

$$1 = e_1 + e_2 + \ldots + e_k \quad \mathrm{mod}(m), \tag{3}$$

$$e_i e_j \equiv 0 \quad \mathrm{mod}(m) \quad \text{for } i \neq j. \tag{4}$$

PROOF   Since p_i, m/p_i are relatively prime, the congruence

$$(m/p_i)x_i \equiv 1 \quad \mathrm{mod}(p_i)$$

is solvable. For a solution x_i denote the left-hand side by e_i. Then (1) and (2) hold and also (2′). But (1) and (2′) imply that the congruences (3), (4) hold modulo all the (p_i) and so also modulo (m) by (i).

(iv) *The 2^k partial sums of $e_1 + e_2 + \ldots + e_k$ are modulo (m) idempotent and no two are congruent modulo (m). Every modulo (m) idempotent element is congruent modulo (m) to one of these partial sums.*

PROOF   Let I be a subset of $\{1, 2, \ldots, k\}$. Then because of (1), (2′)

$$\sum_{j \in I} e_j \equiv \begin{cases} 1 & \mathrm{mod}(p_i) \quad \text{if } i \in I, \\ 0 & \mathrm{mod}(p_i) \quad \text{if } i \notin I. \end{cases}$$

Accordingly the partial sums of $e_1 + e_2 + \ldots + e_k$ are modulo (m) idempotent by (ii). Any two different partial sums are incongruent modulo (m): for if e_1, for example, appears in one of them and not in the other, then the first is congruent to $1 \bmod(p_1)$ and the second is congruent to $0 \bmod(p_1)$. If e is an arbitrary modulo (m) idempotent element, then by (ii) there is a subset I of $\{1, 2, \ldots, k\}$ so that $e \equiv 1$ $\mathrm{mod}(p_i)$ if $i \in I$, and $e \equiv 0 \bmod(p_i)$ if $i \notin I$. Then $e \equiv \sum_{j \in I} e_j$ modulo each (p_i), and so modulo (m).

Thus we have proved

THEOREM 1   *Let R be a principal ideal domain and m a square-free element of R of the form (*). Then there exist 2^k modulo (m) idempotent elements in R with the property that every modulo (m) idempotent element is congruent modulo (m) to exactly one of these elements.*

THEOREM 2   *Let R be a principal ideal domain and m a square-free element of R. Then every element in R is an associate modulo (m) of a modulo (m) idempotent element.*

More explicitly, we assert that *for every $a \in R$ there is a modulo (m) idempotent element $e \in R$ and an element u with the two following properties:*

$$a \equiv ue' \quad \mathrm{mod}(m), \quad \text{with } e' = 1 - e, \tag{5}$$

the congruence $ux \equiv 1 \bmod(m)$ *is solvable.* (6)

PROOF   Again let m be of the form (*) and let $e_1, e_2, \ldots, e_k$ be the modulo (m) idempotent elements from the Chinese construction. Partition the prime elements

$p_1, p_2, \ldots, p_k$ into those which divide a and those which do not divide a. Let $I = \{i : p_i \mid a\}$. Set

$$\sum_{j \in I} e_j = e.$$

Then e is modulo (m) idempotent. For as in the proof of (iv)

$$e \equiv \begin{cases} 1 & \mod(p_i) \quad \text{if } p_i \mid a, \\ 0 & \mod(p_i) \quad \text{if } p_i \nmid a. \end{cases}$$

Thus it follows that $ae \equiv 0$ modulo all (p_i). (If $p_i \mid a$, the first factor $\equiv 0 \mod(p_i)$; if $p_i \nmid a$, the second factor $\equiv 0 \mod(p_i)$.) Thus, $ae \equiv 0 \mod(m)$, and so $a \equiv ae'$ $\mod(m)$, and so also $a \equiv (a+e)e' \mod(m)$; the latter follows because $ee' \equiv 0$ $\mod(m)$. Now set $a+e = u$. Then (5) holds. Moreover,

$$u = a+e \equiv \begin{cases} 0+1 \equiv 1 \not\equiv 0 & \mod(p_i) \quad \text{if } p_i \mid a, \\ a+0 \equiv a \not\equiv 0 & \mod(p_i) \quad \text{if } p_i \nmid a. \end{cases}$$

Thus none of the prime elements $p_1, p_2, \ldots, p_k$ divides u. Accordingly u, m are relatively prime and (6) holds.

REMARK Naturally we could arrange matters so that the modulo (m) idempotent element which is associated modulo (m) with a is called e and not e'. The above notation was chosen because the prime element p_i is an associate modulo (m) of e_i', as the proof of theorem 2 shows.

2
MAIN THEOREMS ABOUT CYCLIC MAPPINGS AND CYCLIC CLASSES

Returning to the theory of n-gons, let us assume the data of §1.1. Over the field K we form the *polynomial ring* $K[x]$. It is a principal ideal domain and also an algebra over K. If we replace the indeterminant x in the polynomials of $K[x]$ by the cyclic mapping ζ, we obtain the cyclic mappings, i.e. the elements of $K[\zeta]$, the (commutative) algebra of cyclic mappings. This replacement by ζ is a homomorphism of $K[x]$ onto $K[\zeta]$. Therefore, by chapter 2, theorem 4,

$$f(x) \equiv 0 \mod(x^n - 1) \text{ is equivalent to } f(\zeta) = 0.$$

Congruences modulo $(x^n - 1)$ in $K[x]$ are equivalent to equations in $K[\zeta]$:

$$f(x) \equiv g(x) \mod(x^n - 1) \text{ is equivalent to } f(\zeta) = g(\zeta).$$

THEOREM 3 $x^n - 1$ *is square-free in* $K[x]$.

PROOF The polynomial $x^n - 1$ is neither the zero polynomial nor a unit of $K[x]$. If it is not square-free, then it has a factor in common with its derivative nx^{n-1} in $K[x]$, and this factor is not a unit. This is the case only when nx^{n-1} is the zero polynomial, that is, if, in K, $n \cdot 1 = 0$. Since by our basic assumption (§1.1) the characteristic of K is not a divisor of n, $n \cdot 1 \neq 0$.

Let k be the number of prime factors of $x^n - 1$ in $K[x]$, and let

$$x^n - 1 = p_1(x)p_2(x) \dots p_k(x) \tag{7}$$

be the prime factorization of $x^n - 1$ in $K[x]$. By Theorem 3 no two of the prime factors $p_i(x)$ are associated.

For a divisor d of n, let $F_d(x)$ be the dth cyclotomic polynomial of $K[x]$ (see appendix 2). Then $x^n - 1$ possesses in $K[x]$ at least the factorization

$$x^n - 1 = \prod_{d \mid n} F_d(x). \tag{8}$$

Therefore k is at least equal to the number $\tau(n)$ of divisors of n. On the other hand, k is at most n, and so

$$\tau(n) \leq k \leq n. \tag{9}$$

If K is the field Q of rational numbers, then it is well known that the cyclotomic polynomials are irreducible, and (8) is thus the prime factorization of $x^n - 1$, and hence $\tau(n) = k$. The case $n = k$ appears when the field K contains the nth roots of unity. $x - 1$ is a divisor of $x^n - 1$ independently of K or n.

If in the modulo $x^n - 1$ idempotent polynomials of $K[x]$ we replace the indeterminant x by the cyclic mapping ζ, we obtain the cyclic projections. Then by theorem 3 the following follows from theorems 1 and 2:

THEOREM 4   *There are exactly 2^k cyclic projections. Every cyclic mapping is an associate of a cyclic projection: given any cyclic mapping φ there exists a cyclic projection π and an invertible cyclic mapping χ with $\varphi = \chi\pi$.*

By the lemma the cyclic projection associated with a cyclic mapping is uniquely determined; we denote it by π_φ. By theorem 4 the set of all cyclic mappings is partitioned into finitely many classes of associates. We can formulate theorem 4 as follows:

THEOREM 4'   *$K[\zeta]$ consists of 2^k classes of associates, and the cyclic projections form a system of representatives of these classes.*

It is now a matter of drawing consequences for the cyclic n-gonal classes from this fact. For this purpose we prove

THEOREM 5   *For cyclic mappings φ, ψ the following assertions are equivalent:*

1. $\varphi \sim \psi$,
2. $(\varphi) = (\psi)$,
3. $\operatorname{Ker} \varphi = \operatorname{Ker} \psi$,
4. $\operatorname{Im} \varphi = \operatorname{Im} \psi$.

In particular, for every cyclic mapping φ

$$\text{Ker } \varphi = \text{Ker } \pi_\varphi, \tag{10}$$

$$\text{Im } \varphi = \text{Im } \pi_\varphi. \tag{11}$$

PROOF (φ) $\subseteq$ (ψ) implies that Ker $\varphi \supseteq$ Ker ψ and Im $\varphi \subseteq$ Im ψ (for $\varphi = \chi\psi$ implies that Ker $\varphi =$ Ker $\chi\psi \supseteq$ Ker ψ and Im $\varphi =$ Im $\chi\psi =$ Im $\psi\chi \subseteq$ Im ψ); (φ) $\supseteq$ (ψ) gives the reversed inclusions. Thus 2 implies 3 and 4.

The special case $\psi = \pi_\varphi$ yields the equations (10) and (11), since (i) $\varphi \sim \pi_\varphi$, and so (ii) (φ) $= (\pi_\varphi)$.

Consider now the statements

1′. $\pi_\varphi \sim \pi_\psi$,
2′. $(\pi_\varphi) = (\pi_\psi)$,
3′. Ker $\pi_\varphi =$ Ker π_ψ,
4′. Im $\pi_\varphi =$ Im π_ψ.

By (i), (ii), (10), (11) and the corresponding expressions for ψ, we have the following equivalences: 1 and 1′; 2 and 2′; 3 and 3′; 4 and 4′. Each of the statements 1′–4′ is equivalent to $\pi_\varphi = \pi_\psi$. This is true for 1′ and 2′ by the lemma, and for 4′ by chapter 2, theorem 7. By chapter 2, (7), 3′ is equivalent to $\text{Im}(1-\pi_\varphi) = \text{Im}(1-\pi_\psi)$, and thus, by the same theorem 7, equivalent to $1-\pi_\varphi = 1-\pi_\psi$, and so to $\pi_\varphi = \pi_\psi$.

As a consequence of the second statement of theorem 4 and of theorem 5 we have

THEOREM 6 *Every cyclic mapping is a quasi-projection.*

FIRST PROOF In every commutative ring with 1 the following holds: $a \sim b$ implies $a^2 \sim b^2$. If a is an associate of an idempotent element e, then $a^2 \sim e$, thus $a^2 \sim a$. Thus for every cyclic mapping, by theorem 4 $\varphi^2 \sim \varphi$, and so by theorem 5, Im $\varphi^2 =$ Im φ and Ker $\varphi^2 =$ Ker φ. Therefore φ is a quasi-projection.

SECOND PROOF Let φ by a cyclic mapping. Since by chapter 2, theorem 8, $\mathfrak{A}_n = \text{Im } \pi_\varphi \oplus \text{Ker } \pi_\varphi$, then by (10) and (11), $\mathfrak{A}_n = \text{Im } \varphi \oplus \text{Ker } \varphi$. Then by chapter 2, theorem 8′, φ is a quasi-projection of $\mathfrak{A}_n$.

The cyclic classes of n-gons are the kernels of the cyclic mappings (chapter 2, theorem 1). Since two cyclic mappings have the same kernel if and only if they are associates (theorem 5), and since by theorem 4′ there are exactly 2^k classes of associated cyclic mappings, there are exactly 2^k cyclic classes of n-gons.

The fact that for a given n there are only finitely many cyclic n-gonal classes (to be precise, 2^k) is the most tangible part of the statement of the main theorem on cyclic classes.

Let us obtain further information about cyclic classes by proceeding a bit differently and restricting ourselves more closely to the properties of cyclic projections.

Every cyclic mapping φ has the same image and kernel as the cyclic projection π_φ (by (10) and (11)). Thus for any φ there exists a cyclic projection ε with Ker φ = Im ε, Im φ = Ker ε. For, denoting by ε_φ the complement of π_φ, i.e. the cyclic projection $1 - \pi_\varphi$, we have (by (7) of chapter 2)

$$\text{Ker } \varphi = \text{Ker } \pi_\varphi = \text{Im } \varepsilon_\varphi, \tag{12}$$

$$\text{Im } \varphi = \text{Im } \pi_\varphi = \text{Ker } \varepsilon_\varphi. \tag{13}$$

Trivially the π_φ are all the cyclic projections (for every cyclic projection is associated with itself), and the same is true of their complements ε_φ. Thus from (12) and (13) we have: *The following sets coincide*:

1. *the kernels of the cyclic mappings,*
2. *the kernels of the cyclic projections,*
3. *the images of $\mathfrak{A}_n$ under the cyclic projections,*
4. *the images of $\mathfrak{A}_n$ under the cyclic mappings.*

Since the kernels of the cyclic mappings are the cyclic classes (chapter 2, theorem 1), we have, accordingly, four characterizations of cyclic classes. Stressing 3, we have (see (12))

THEOREM 7   *Every cyclic class is the image of $\mathfrak{A}_n$ under a cyclic projection.*

At this point all questions about cyclic classes and cyclic mappings posed in chapter 2 have been answered.

The Boolean algebra $E(K[\zeta])$, $\circ$, $\cdot$ of cyclic projections (see §5.4) contains 0 and 1 and by theorem 4 has exactly 2^k elements. It is a Boolean algebra of endomorphisms of $\mathfrak{A}_n$. Applying Im-transfer to $E(K[\zeta])$ (chapter 5, theorem 3), we obtain a Boolean algebra of 2^k cyclic classes (chapter 2, theorem 9), and by theorem 7 this Boolean algebra contains all cyclic classes. Thus:

MAIN THEOREM   *The cyclic n-gonal classes form a finite Boolean algebra.*
The Boolean algebra of cyclic n-gonal classes is a sublattice of the lattice of subspaces of the vector space $\mathfrak{A}_n$. If k is the number of prime factors of $x^n - 1$ in $K[x]$, then the number of cyclic n-gonal classes is 2^k.

No matter how 'immensely infinite' the lattice of subspaces of $\mathfrak{A}_n$ may be, the concept of cyclic n-gonal classes chooses a finite Boolean net from it.

COROLLARIES OF THE MAIN THEOREM
The sum and intersection of two cyclic n-gonal classes are cyclic n-gonal classes.

$\mathfrak{A}_n$ is the direct sum of the atomic cyclic n-gonal classes, i.e. of atoms of the Boolean algebra of cyclic n-gonal classes.

Every n-gon is uniquely representable as a sum of n-gons from the atomic cyclic classes.

The concept of atomic cyclic n-gonal classes, as well as the number k of these classes, depends upon the field K. We shall study this more closely in chapters 10–12. Roughly speaking, the n-gons of these classes are distinguished by regularity.

EXAMPLE $n = 6$, $K = Q$. Since $\tau(6) = 4$, there are $2^4 = 16$ cyclic classes; the 16 cyclic hexagonal classes in §5.5 are thus all the cyclic hexagonal classes, and the decomposition of a hexagon given there is the atomic decomposition.

If, using the Chinese construction, we determine for every prime factor $p_i(x)$ of $x^n - 1$ a polynomial $e_i(x) \in K[x]$ with

$$e_i(x) \equiv 1 \quad \mathrm{mod}(p_i(x)), \tag{14}$$

$$e_i(x) \equiv 0 \quad \mathrm{mod}\left(\frac{x^n-1}{p_i(x)}\right), \tag{15}$$

then $e_1(\zeta), e_2(\zeta), \ldots, e_k(\zeta)$ are mutually orthogonal cyclic projections $\neq 0$ with sum 1; they are the atomic cyclic projections, i.e. the atoms of $E(K[\zeta])$ (see chapter 5, theorem 2). Im $e_1(\zeta)$, Im $e_2(\zeta)$, ..., Im $e_k(\zeta)$ are the atomic cyclic n-gonal classes, and

$$A = e_1(\zeta)A + e_2(\zeta)A + \ldots + e_k(\zeta)A \tag{16}$$

is the decomposition of an n-gon A into its *atomic components*.

The n-gons for which prescribed atomic components are zero form a cyclic class, and each of the 2^k cyclic n-gonal classes arises in this way. Thus for $K = Q$ the hexagons whose affinely regular components are zero are the prisms (cf. §5.5).

Exercises

1 Let φ be a cyclic mapping. If $\mathbb{C}$ is a cyclic class, then $\varphi\mathbb{C}$ is a cyclic class with $\varphi\mathbb{C} \subseteq \mathbb{C}$ (this is a refinement of chapter 2, theorem 5). In fact, $\varphi\mathbb{C} = \pi_\varphi\mathbb{C}$. In the Boolean algebra of cyclic classes consider the intervals [Ker φ, $\mathfrak{A}_n$] and [{0}, Im φ]. If $\mathbb{C}$ runs through the elements of the first interval, $\mathbb{C} \to \varphi\mathbb{C}$ is an isomorphism of the first interval onto the second (see figure 57). Which cyclic mappings have the property that every cyclic class belongs to exactly one of the two intervals? (See appendix 1, §2.5, §3.2.)

2 Let the involutory cyclic mappings be called *cyclic reflections*. They form a subgroup of the group of units of $K[\zeta]$ and have coefficient sum ± 1. Now let char $K \neq 2$. With respect to the composition defined in §5.1, exercise 2, the cyclic reflections

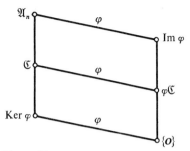

Figure 57

also form a Boolean algebra. $\varepsilon \to 2\varepsilon - 1$ is an isomorphism of the Boolean algebra of cyclic projections onto the Boolean algebra of cyclic reflections. We have Fix ε = Fix$(2\varepsilon - 1)$, Track ε = Track$(2\varepsilon - 1)$.[2] There are exactly 2^k cyclic reflections.

$$\varphi \to \text{Fix } \varphi$$

is an isomorphism of the Boolean algebra of cyclic reflections onto the Boolean algebra of cyclic classes. The cyclic reflection φ with Fix $\varphi = \mathfrak{C}$ is the 'reflection in $\mathfrak{C}$', i.e. the endomorphism of V^n which keeps fixed each element of $\mathfrak{C}$ and maps every n-gon A of the class complementary to $\mathfrak{C}$ onto $-A$. (If $\varphi = 2\varepsilon - 1$, we can also say that under φ every n-gon is reflected onto its ε-image.)

In the group of cyclic reflections the isobaric ones form a subgroup of index 2; the non-isobaric ones are obtained from them by multiplying by -1.

An infimum of distinct anti-atoms of the Boolean algebra of cyclic reflections is equal to the product of these anti-atoms. The cyclic reflections are the partial products of the product of all the anti-atoms.

3 If $n = 2m$, ζ^m is an isobaric cyclic reflection. Let $K = Q$. If $n = 4$, the isobaric cyclic reflections form a Klein four-group, consisting of 1, ζ^2, $\frac{1}{2}(-1 + \zeta + \zeta^2 + \zeta^3)$, $\frac{1}{2}(1 + \zeta + \zeta^2 + \zeta^3)$. Determine the isobaric cyclic reflections for $n = 6$.

3

THE PRIME FACTORS OF $x^n - 1$ AND THE ATOMIC CYCLIC CLASSES

For every prime factor $p_i(x)$ of $x^n - 1$ we form the cyclic mapping $p_i(\zeta)$. The cyclic projection associated with it is $1 - e_i(\zeta)$ (this follows from the remark at the end of §1). Thus by (12)

$$\text{Ker } p_i(\zeta) = \text{Im } e_i(\zeta). \tag{17}$$

The kernels of the cyclic mappings $p_i(\zeta)$ are the atomic cyclic classes.

Since generally the kernel of a cyclic mapping $c_0 + c_1\zeta + \ldots + c_{n-1}\zeta^{n-1}$ is the

---

2   Track ψ is defined as Im$(1 - \psi)$.

solution space of the cyclic system of equations $c_0 a_1 + c_1 a_2 + \ldots + c_{n-1} a_n = o, \ldots$ (see §2.2), we can also express this fact as follows:

THEOREM 8 *The cyclic systems of equations whose coefficient n-tuples are the coefficient n-tuples of the prime factors of $x^n - 1$ in $K[x]$ describe the atomic cyclic n-gonal classes.*

In this connection, by *coefficient n-tuple of a proper divisor* $\sum c_i x^i$ of $x^n - 1$ we understand the n-tuple $(c_0, \ldots, c_{n-1})$. The coefficient n-tuple of $x^n - 1$ should be $(0, 0, \ldots, 0)$. This convention is, for the moment, meaningful only for the case $n = 1$, which presents no problem.

One atomic cyclic class is the class of trivial n-gons; all other atomic cyclic classes are zero-point classes (cf. the note about addition of n-gons).

The cyclic system of equations associated with the prime divisor $x - 1$ describes the class of trivial n-gons. (If $n \neq 1$, $x - 1$ has coefficient n-tuple $(-1, 1, 0, \ldots, 0)$; the cyclic system of equations belonging to it is $-a_1 + a_2 = o, \ldots$ and defines the class of trivial n-gons. If $n = 1$, we have a special case: $x - 1$ has coefficient 1-tuple (0); the cyclic system of equations belonging to it is $0 a_1 = o$ and defines the class of all 1-gons which is also the class of trivial 1-gons.)

If we now set $p_1(x) = x - 1$, then the cyclic systems of equations whose coefficient n-tuples are the coefficient n-tuples of the prime factors $p_i(x)$ with $i \neq 1$ define the atomic zero-point classes.

We can also see that these systems of equations define zero-point classes as follows: $p_2(x) \ldots p_k(x) = x^{n-1} + \ldots + x + 1$. If we substitute 1 for x we obtain

$$p_2(1) \ldots p_k(1) = n.$$

Since generally $f(1)$ is the coefficient sum of the polynomial $f(x)$, the polynomials $p_i(x)$ with $i \neq 1$ have coefficient sums different from 0, and the cyclic systems of equations formed from their coefficients define zero-point classes (by chapter 1, theorem 1).

The polynomial $p_i(x)$ with $i \neq 1$ can be normalized by multiplication with $1/p_i(1)$, so that its coefficient sum becomes 1. The cyclic mapping arising by substitution of ζ is then isobaric and has the same kernel as $p_i(\zeta)$.

If K is the field of rational numbers Q, then the cyclic systems of equations whose coefficient n-tuples are the coefficient n-tuples of the cyclotomic polynomials $F_d(x)$ with $d \mid n$ define the atomic cyclic classes. $F_1(x) = x - 1$.

EXAMPLE $K = Q, n = 6$. Then

$$x_6 - 1 = F_1(x) F_2(x) F_3(x) F_6(x) = (x - 1)(x + 1)(x^2 + x + 1)(x^2 - x + 1)$$

and $F_2(1) = 2$, $F_3(1) = 3$, $F_6(1) = 1$. The coefficient n-tuples of the polynomials $F_d(x)$ (with $d = 2, 3, 6$), normalized to have coefficient sum 1, yield the cyclic systems of equations

$d = 2$: $\frac{1}{2}(a_1 + a_2) = o, \ldots$;
$d = 3$: $\frac{1}{3}(a_1 + a_2 + a_3) = o, \ldots$;
$d = 6$: $a_1 - a_2 + a_3 = o, \ldots$;

and obviously these cyclic systems of equations define the class $\mathfrak{A}_{2,3}$ of triply counted digons with centre of gravity o, the class $\mathfrak{A}_{3,2}$ of doubly counted triangles with centre of gravity o, and the class $\mathfrak{R}_6$ of affinely regular hexagons with centre of gravity o. Thus, together with the class of trivial hexagons, these are the atomic cyclic classes. The cyclic mappings obtained from the normalized polynomials $F_d(x)$ with $d = 2, 3, 6$ by substitution of ζ, and whose kernels were determined just above, are

$$\tfrac{1}{2}F_2(\zeta) = \kappa_2, \qquad \tfrac{1}{3}F_3(\zeta) = \kappa_3, \qquad F_6(\zeta) = \alpha_3,$$

i.e. precisely those mappings which we encountered in chapter 2 in connection with the hexagonal classes.

Exercises

1 For every n and every allowable K, the centre-of-gravity n-gon of A is an atomic component of A.

2 Let $n = 4$ and -1 be a non-square in K. Let A be a quadrangle, and set $A - \sigma A = A_0$ (cf. §2.4). The atomic components of A are σA, the doubly counted digon formed by the mid-points of the diagonals of A_0, and the parallelogram which has the same parallelogram of mid-edge points as A_0. If -1 is a square in K, then this parallelogram is decomposed further into two squares (see chapter 11).

PART III
Boolean algebras of n-gonal theory

The considerations of §6.1 and Im-transfer, which formed the algebraic background for the proof of the main theorem about cyclic n-gonal classes, will now be illuminated from a more systematic point of view. We shall use this point of view in the further development of the theory of n-gons.

Chapter 7 is purely algebraic in nature. In it we shall demonstrate, for a principal ideal domain R and an element $m \in R$, the relations between the divisors of m, the ideals of R generated by them, the ideals of the residue class ring $R/(m)$, and the idempotent elements of $R/(m)$. The case where m is square-free is of particular interest.

In chapter 8 these considerations will be applied to the theory of n-gons; in particular, R will be the polynomial ring $K[x]$, m the square-free polynomial $x^n - 1$, and the residue class ring, up to isomorphism, will be the algebra of cyclic mappings of the vector space of n-gons. In §8.2 we shall formulate generally the fundamental connection between the divisors of $x^n - 1$ and the cyclic n-gonal classes, a connection which has already been mentioned in §6.3 for prime divisors of $x^n - 1$ and the atomic cyclic n-gonal classes. This connection will be realized in chapter 8 by means of cyclic projections (idempotent cyclic mappings), as has already been done in the partial statement of §6.3 (see figure 60).

At the beginning of chapter 7 we give a new conception of the theorem of Im-transfer by writing it as a theorem about R-modules. This general theorem of idempotent-transfer has a further special case which is of importance for us (§7.3).

In contrast to the idempotent-transfer theorem, in chapter 9 we shall give a second transfer theorem about R-modules, the ideal-transfer theorem. It describes a general situation from which we can obtain, by specialization, the main theorem on cyclic n-gonal classes. It will show us how to prove the main theorem and obtain, without considering cyclic projections, the connection between the divisors of $x^n - 1$ and cyclic n-gonal classes.

7
Idempotent-transfer. Residue class rings of principal ideal domains

1
R-MODULES

Let $R, +, \cdot$ be a ring with 1, and $\mathfrak{A}, +$ an abelian group. The elements of R are denoted by $a, b, \ldots, r, s, \ldots$ and the elements of $\mathfrak{A}$ by $\boldsymbol{a}, \boldsymbol{b}, \ldots$. With respect to a mapping

$$(r, \boldsymbol{a}) \to r\boldsymbol{a} \tag{1}$$

of $R \times \mathfrak{A}$ into $\mathfrak{A}$ as multiplication, $\mathfrak{A}$ is called an *R-module* if the following rules hold:

(i) $r(\boldsymbol{a}+\boldsymbol{b}) = r\boldsymbol{a}+r\boldsymbol{b}$,
(ii) $(r+s)\boldsymbol{a} = r\boldsymbol{a}+s\boldsymbol{a}$,
(iii) $(rs)\boldsymbol{a} = r(s\boldsymbol{a})$,
(iv) $1\boldsymbol{a} = \boldsymbol{a}$.

For every $r \in R$, $r\mathfrak{A}$ is a subgroup of $\mathfrak{A}$. A binary relation between elements of R and elements of $\mathfrak{A}$ is defined by

$$r\boldsymbol{a} = \boldsymbol{o} \tag{2}$$

and we read this as r *annihilates* $\boldsymbol{a}$. The annihilator of $\mathfrak{A}$, written annih $\mathfrak{A}$, is defined by

$$\text{annih } \mathfrak{A} = \{r : r\boldsymbol{a} = \boldsymbol{o} \text{ for all } \boldsymbol{a}\} = \{r : r \mathfrak{A} = \{\boldsymbol{o}\}\}.$$

annih $\mathfrak{A}$ is a (two-sided) ideal of R. (Actually, it might be clearer to write $\text{annih}_R \mathfrak{A}$.)

EXAMPLE 1   Let $R = \text{End}(\mathfrak{A})$ and (1) be the operation of an endomorphism on an element of $\mathfrak{A}$. Then annih $\mathfrak{A} = (0)$.

EXAMPLE 2 $R, +$ is an R-module if we take multiplication in R for (1). Since 1 is annihilated only by 0, annih $R = (0)$.

2
IDEMPOTENT-TRANSFER

Now let R be commutative. Then each $r\mathfrak{A}$ is an R-submodule of $\mathfrak{A}$. Let $E(R)$, $\circ$, $\cdot$ be the Boolean algebra of idempotent elements of R (chapter 5, theorem 1). For elements $e, f \in E(R)$ we have

$$e\mathfrak{A} = f\mathfrak{A} \quad \text{implies} \quad e \equiv f \mod \text{annih } \mathfrak{A}, \tag{3}$$

$$(e \circ f)\mathfrak{A} = e\mathfrak{A} + f\mathfrak{A}, \tag{4}$$

$$ef\mathfrak{A} = e\mathfrak{A} \cap f\mathfrak{A}. \tag{5}$$

PROOF   (3) is proved just as is theorem 7 of chapter 2, while (4) and (5) are proved in the same way as the rules (15), (16) in §5.3. Let $L(\mathfrak{A})$ denote the lattice of R-submodules of $\mathfrak{A}$. Then from (3)–(5) we obtain a new formulation of theorem 3 of chapter 5:

THEOREM 1 (Idempotent-Transfer)   *Let R be a commutative ring with 1, $\mathfrak{A}$ an R-module*, annih $\mathfrak{A} = (0)$. *Then*

$$e \to e\mathfrak{A} \tag{6}$$

is an isomorphism of the Boolean algebra $E(R)$, $\circ$, $\cdot$ onto a sublattice of $L(\mathfrak{A})$, $+$, $\cap$.

This transfer yields a Boolean algebra of R-submodules of $\mathfrak{A}$.

3
A SPECIAL CASE OF IDEMPOTENT-TRANSFER

Once more let R be commutative. Think of R as an R-module (example 2). The R-submodules of R are the ideals of R; in particular, rR is the principal ideal (r) generated by r. Let us denote the lattice of ideals of R by $L(R)$. Then by theorem 1

$$e \to eR = (e) \tag{7}$$

is an isomorphism of the Boolean algebra $E(R)$, $\circ$, $\cdot$ onto a sublattice of the lattice of ideals $L(R)$, $+$, $\cap$.

The connection between idempotents and ideals of R is especially simple if R is a direct sum of fields $K_1, K_2, \ldots, K_k$:

$$R = K_1 \oplus K_2 \oplus \ldots \oplus K_k.$$

For in R we have a representation of the unit element $1_R = (1, 1, \ldots, 1)$ as the sum

$$1_R = e_1 + e_2 + \ldots + e_k \tag{8}$$

of mutually orthogonal idempotent elements

$$e_1 = (1, 0, ..., 0), \quad e_2 = (0, 1, 0, ..., 0), \quad ..., \quad e_k = (0, ..., 0, 1). \tag{9}$$

$E(R)$ consists of the partial sums of $e_1 + e_2 + ... + e_k$, and e_1, e_2, ..., e_k are the atoms of $E(R)$ (see §5.1, example 3, and chapter 5, theorem 2).

The transfer (7) carries (8) into

$$R = (e_1) + (e_2) + \ ... \ + (e_k) \tag{10}$$

and every partial sum of $e_1 + e_2 + ... + e_k$ into the corresponding partial sum of $(e_1) + (e_2) + ... + (e_k)$. (e_i) consists of all the ring elements $(0, ..., 0, a_i, 0, ..., 0)$ and is isomorphic to K_i.

Besides this, we have: *every ideal M of R is a partial sum of* $(e_1) + (e_2) + ... + (e_k)$.

PROOF　$M = e_1 M + e_2 M + ... + e_k M$ ($\subseteq$ follows from $M = 1_R \cdot M$ and (8); $\supseteq$ follows from $M \supseteq e_i M$), and every ideal $e_i M$ is either the zero ideal (0) or equal to (e_i) (for $(e_i) \supseteq e_i M$, and, since (e_i) is isomorphic to a field, (e_i) contains only the ideals (0) and (e_i)).

THEOREM 2　*If R is a (finite) direct sum of fields, then* (7) *is an isomorphism of E(R) onto the (full) lattice of ideals of R.*

Thus in a direct sum of k fields every ideal is a principal ideal generated by an idempotent element, and the lattice of ideals is a Boolean algebra with 2^k elements.

4
IDEALS AND DIVISIBILITY IN A PRINCIPAL IDEAL DOMAIN

Let R be a principal ideal domain. $(r) \supseteq (s)$ is equivalent to 'r divides s' or $r \mid s$. If $r \mid s$, then every associate of r divides every associate of s. (1) $(r) = (s)$ is equivalent to (2) $r \mid s$ and $s \mid r$. Since R is an integral domain, this is equivalent to (3) $r \sim s$.

A g.c.d. of a, b is a generator of the sum ideal $(a) + (b) = (a, b)$, and a l.c.m. of a, b is a generator of the intersection ideal $(a) \cap (b)$. The g.c.d. and l.c.m. of two elements of R are determined up to associates.

The relation 'divides' and the concepts of g.c.d. and l.c.m. can be taken over to classes of associates. *With respect to 'divides' as a partial ordering, the classes of associates form a lattice L, with l.c.m. and g.c.d. as supremum and infimum.* The mapping $r \to (r)$ gives an anti-isomorphism of the lattice L onto the lattice $L(R)$ of ideals of R.

In a principal ideal domain we have unique prime factorization (a more precise formulation is given in §6.1). Some important properties of principal ideal domains in connection with this are: every properly ascending chain of ideals is finite; the lattices L, $L(R)$ are distributive (cf. appendix 1).

If m is an element of R, then the classes of associated divisors of m form a sublattice $L(m)$ of the lattice L. $L(m)$ *is called the lattice of divisors of m. L(m) is distributive and, if m $\neq$ 0, finite.*

If m is square-free, that is, can be represented as

$$m = p_1 p_2 \dots p_k \text{ with mutually non-associated prime elements } p_i \ (k \geq 1), \qquad (*)$$

then the 2^k partial products of the product $p_1 p_2 \dots p_k$ are non-associated and represent the divisors of m exactly (the empty partial product is set equal to 1). The lattice of divisors of m is complemented, and so is a Boolean algebra with 2^k elements; the prime elements $p_1, p_2, \dots, p_k$ represent the atoms.

An ideal (m) is called *square-free* if m is square-free.

5

RESIDUE CLASS RINGS OF PRINCIPAL IDEAL DOMAINS

Once more let R be a principal ideal domain and $m \neq 0$ an element of R. We shall consider the residue class ring $R/(m)$ and the following lattices:

L_1: *The lattice of divisors of* m;
L_2: *The interval* $[(m), R]$ *of the lattice of ideals of* R;
L_3: *The lattice of ideals of* $R/(m)$;
L_4: $E(R/(m))$, $\circ$, $\cdot$, *the Boolean algebra of idempotents from* $R/(m)$.

Let divisors of m be denoted by t. Then the mapping

$$t \rightarrow (t)$$

defines an anti-isomorphism i_{12} of L_1 onto L_2.

Further, the canonical homomorphism $a \rightarrow a+(m)$ of R onto the residue class ring $R/(m)$ induces an isomorphism of the lattice of ideals of R lying above the kernel (m) onto the lattice of ideals of $R/(m)$. Since the ideals of R lying between (m) and R are the ideals (t) with $t \mid m$, the mapping

$$i_{23} : (t) \rightarrow (t+(m))$$

is an isomorphism of L_2 onto L_3.

THEOREM 3. *The product of* i_{12}, i_{23} *maps the lattice of divisors of m anti-isomorphically onto the lattice of ideals of* $R/(m)$. *Each ideal in* $R/(m)$ *can be written in the form* $(t+(m))$ *with* $t \mid m$. *If* $t_1, t_2 \mid m$, *then* $(t_1+(m)) = (t_2+(m))$ *if and only if* $(t_1) = (t_2)$.

In particular, theorem 3 says that every ideal in $R/(m)$ is a principal ideal. (Nevertheless $R/(m)$ can have zero divisors, in which case it is not a principal ideal domain, but merely a principal ideal ring.)

Even if a is an arbitrary element of R, $(a+(m))$ is an ideal in $R/(m)$, and we have:

LEMMA   (i) *If* $(a, m) = (t)$, *then* $(a+(m)) = (t+(m))$; (ii) $(a_1+(m)) = (a_2+(m))$ *is equivalent to* $(a_1, m) = (a_2, m)$.

PROOF OF (i)   If $a \in (t)$, then $a+(m) \in (t+(m))$. On the other hand, since $t \in (a, m)$, t has a representation $t = ua+vm$ with $u, v \in R$. Therefore $t+(m) = ua+vm+(m) = ua+(m) \in (a+(m))$.

PROOF OF (ii)   Let $(a_i, m) = (t_i)$. Then by (i), $(a_i+(m)) = (t_i+(m))$ for $i = 1, 2$. By theorem 3, $(t_1+(m)) = (t_2+(m))$ is equivalent to $(t_1) = (t_2)$.

An element $e \in R$ is modulo (m) idempotent if and only if the residue class $e+(m)$ is an idempotent element of $R/(m)$. By §3

$$i_{43} : e+(m) \to (e+(m))$$

is an isomorphism of L_4 onto a Boolean subalgebra of L_3.

The product of i_{12}, i_{23} gives an approach to all ideals of $R/(m)$, and i_{43} gives an approach to at least some ideals. The question whether we can find for a given ideal of $R/(m)$, written as $(t+(m))$ with $t \mid m$, an idempotent residue class which also generates the ideal will be answered by the construction of the 'Chinese remainder theorem' (see theorem 4').

THEOREM 4 (Chinese Remainder Theorem)   *In a principal ideal domain R, let an element $m \neq 0$ be represented as a product of relatively prime elements: $m = t_1 t_2 \dots t_k$. Then there exist residue classes*

$$e_1+(m), e_2+(m), \dots, e_k+(m) \tag{11}$$

with the following properties:

(i) *The residue classes (11) are mutually orthogonal, and their sum is the residue class of 1.*
(ii) $(t_i+(m)) = (1-e_i+(m))$.

PROOF   Set $m/t_i = \hat{t}_i$. Since $t_i, \hat{t}_i$ are relatively prime, the congruence

$$\hat{t}_i x_i \equiv 1 \quad \mathrm{mod}(t_i)$$

is solvable in R. For a solution x_i denote the left side by e_i. Then

$$e_i \equiv 1 \quad \mathrm{mod}(t_i), \tag{12}$$

$$e_i \equiv 0 \quad \mathrm{mod}(\hat{t}_i). \tag{13}$$

Thus $e_i \equiv 0 \, \mathrm{mod}(t_j)$ for $i \neq j$, and moreover

$$1 \equiv e_1+e_2+\dots+e_k \quad \mathrm{mod}(m), \tag{14}$$

$$e_i e_j \equiv 0 \quad \mathrm{mod}(m) \quad \text{for } i \neq j. \tag{15}$$

For these last two congruences hold modulo each (t_i), and since the t_i's are relatively prime, they also hold modulo (m). Congruences (14) and (15) yield assertion (i) when read as equations in the residue class ring $R/(m)$.

Multiplying (14) by e_i and using (15), we have: the elements e_i are modulo (m) idempotent. Then this also holds for $1 - e_i$.

To prove (ii) we note that, since $1 - e_i \in (t_i)$ by (12) and $m \in (t_i)$, $(1 - e_i, m) \subseteq (t_i)$. On the other hand, by (13) there exists a $u_i \in R$ with $e_i = u_i \bar{t}_i$. Thus $1 = 1 - e_i + u_i \bar{t}_i$, and multiplication by t_i yields $t_i \in (1 - e_i, m)$. Thus $(t_i) = (1 - e_i, m)$, and assertion (ii) then follows from the lemma.

THEOREM 4' (Special Case of the Chinese Remainder Theorem) *In a principal ideal domain R, let m be an element of $\neq 0$, and t a divisor of m which is relatively prime to m/t. Then the congruences*

$$e \equiv 1 \quad \mathrm{mod}(t), \tag{16}$$

$$e \equiv 0 \quad \mathrm{mod}(m/t) \tag{17}$$

are solvable in R (uniquely modulo (m)). For a solution e we have: e, $1 - e$ are modulo (m) idempotent, and

$$(t + (m)) = (1 - e + (m)). \tag{18}$$

COROLLARY *In a residue class ring $R/(m)$ of a principal ideal domain R with respect to a square-free ideal (m), every ideal is a principal ideal generated by an idempotent element.* (Cf. theorem 2 of chapter 6.)

PROOF By theorem 3 every ideal of $R/(m)$ can be written in the form $(t + (m))$ where $t \mid m$. Then theorem 4' shows that the ideal in the case of a square-free m is also generated by an idempotent residue class.

THEOREM 5 *If R is a principal ideal domain and m a square-free element of R of the form (*), then the lattices L_1, L_2, L_3, L_4 are Boolean algebras with 2^k elements. The product of i_{12}, i_{23} maps L_1 anti-isomorphically onto L_3 via L_2, and i_{43} is an isomorphism of L_4 onto L_3.*

PROOF L_1 is a Boolean algebra with 2^k elements (see §4). Since i_{12} is an anti-isomorphism of L_1 onto L_2, and i_{23} an isomorphism of L_2 onto L_3, L_2 and L_3 are also Boolean algebras with 2^k elements. i_{43} is, first of all, an isomorphism of the Boolean algebra L_4 onto a sublattice of L_3, but in the case of a square-free m it maps L_4 onto the full lattice of ideals L_3, by the corollary of theorem 4'. Therefore, L_4 also consists of 2^k elements.

Let us denote a solution of the congruences (16), (17) by e_t. (The e_i of theorem 4 are, in this notation, e_{t_i}.) Then equation (18) gives the

RULE $\quad (t+(m)) = (1-e_t+(m))$.

Since $1-e_t+(m)$, $e_t+(m)$ are complementary elements of L_4, this rule with i_{43} implies that

$(t+(m))$, $(e_t+(m))$ *are complementary ideals of $R/(m)$.*

This notation e_t for a solution of the congruences (16), (17) is justified by

THEOREM 6 (Chinese Isomorphism)   *If m is a square-free element of a principal ideal domain R, then*

$$t \rightarrow e_t+(m) \tag{19}$$

defines an isomorphism of the lattice of divisors of m onto the Boolean algebra of idempotents of the residue class ring $R/(m)$.

PROOF   The product of the anti-isomorphism i_{12} of L_1 onto L_2, the isomorphism i_{23} of L_2 onto L_3, complement formation in L_3 (it is an involutory anti-automorphism of L_3), and the isomorphism i_{43}^{-1} of L_3 onto L_4 is an isomorphism of L_1 onto L_4 and maps as follows:

$$t \rightarrow (t) \rightarrow (t+(m)) \rightarrow (e_t+(m)) \rightarrow e_t+(m).$$

For later use in the theory of n-gons we shall formulate these relations among the Boolean algebras under discussion (for square-free m) somewhat differently.

In order to avoid anti-isomorphisms, let us replace the Boolean algebras L_2, L_3, L_4 by their respective duals $L_2', L_3'. L_4'$. Then, first of all

THEOREM 5'   *Under the assumptions of theorem 5, L_1, L_2', L_3', L_4' are Boolean algebras with 2^k elements; they are connected by the natural isomorphisms i_{12}, i_{23}, i_{43} of L_1 onto L_2', L_2' onto L_3', L_4' onto L_3' respectively.*

Further, let us take, as L_5, a second copy of L_4, $E(R/(m))$, $\circ$, $\cdot$, and consider the Boolean algebras

$$L_1, L_2', L_3', L_4', L_5 \quad (L_5 : E(R/(m)), \circ, \cdot).$$

Complement formation in $E(R/(m))$ is an (involutory) anti-automorphism of this Boolean algebra, thus an isomorphism of L_4' onto L_5, and conversely. The Chinese isomorphism maps L_1 onto L_5. (See figure 58.)

Let us now say something with regard to the interpretation of the above rule. If we associate with each divisor t of m an ideal of $R/(m)$ using, on the one hand, the mapping $t \rightarrow (t) \rightarrow (t+(m))$ and, on the other hand, the mapping $t \rightarrow e_t+(m)$ $\rightarrow 1-e_t+(m) \rightarrow (1-e_t+(m))$, then the single steps of the two mappings are isomorphisms of L_1 onto L_2', L_2' onto L_3' and L_1 onto L_5, L_5 onto L_4', L_4' onto L_3' (respectively). Thus both maps are themselves isomorphisms of L_1 onto L_3'; to any element of L_1 they associate the same element of L_3'.

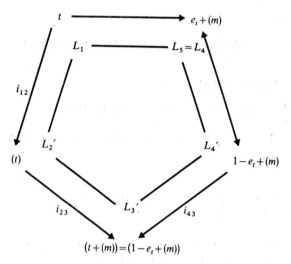

Figure 58

6

RESIDUE CLASS RINGS AS SUMS OF RESIDUE CLASS RINGS

THEOREM 4″ (Enlargement of the Chinese Remainder Theorem) *Under the assumptions of theorem* 4, $R/(m) \cong \sum \oplus R/(t_i)$.

PROOF Let $\varphi_i : a \to a + (t_i)$ be the canonical homomorphism of R onto $R/(t_i)$. Then

$$\varphi : a \to (\varphi_1 a, \varphi_2 a, ..., \varphi_k a)$$

is a homomorphism of R into the direct sum of the $R/(t_i)$'s. The kernel of φ is (m) because a lies in the kernel of φ if and only if $a \equiv 0$ modulo each (t_i), and so modulo (m). If elements $a_1, a_2, ..., a_k \in R$ are given, and if $e_1, e_2, ..., e_k$ are the elements of R constructed in the proof of theorem 4(i), then by (12) and (13)

$$a_1 e_1 + a_2 e_2 + ... + a_k e_k \equiv a_i \quad \mathrm{mod}(t_i) \quad \text{for } i = 1, 2, ..., k.$$

Therefore φ is a mapping of R *onto* the direct sum, and the homomorphism theorem for rings gives the assertion.

Expressed more precisely, we have

$$R/(m) = \sum \oplus (e_i + (m)) \tag{20}$$

and

$$(e_i + (m)) \cong R/(t_i). \tag{21}$$

For because of theorem 4(i) the partial sums of the sum of all residue classes (11) form a Boolean subalgebra of the Boolean algebra L_4, by corollary 1 to chapter 5, theorem 2. The subalgebra can be transferred by i_{43} (idempotent-transfer) into the lattice of ideals L_3; we obtain (20). The isomorphism of $R/(m)$ onto the direct sum of the $R/(t_i)$'s, induced by φ, maps, because of (12), (13), $ae_i + (m)$ onto $(0, ..., 0, \varphi_i a, 0, ..., 0)$; therefore (21) holds for $\{ae_i + (m) : a \in R\} = (e_i + (m))$.

COROLLARY TO THEOREM 4" *If m is square-free and of the form (*), then $R/(m) \cong \sum \oplus R/(p_i)$ and the $R/(p_i)$ are fields.*

This follows from the fact that the factors in (*) are relatively prime and that in a principal ideal domain R each ideal generated by a prime is maximal; the latter means that the residue class ring of R with respect to such an ideal is a field.

Hence the residue class ring of a principal ideal domain with respect to a square-free ideal is a direct sum of fields. Thus for a residue class ring in this case it follows immediately from theorem 2 that its lattice of ideals is a Boolean algebra isomorphic to the Boolean algebra of its idempotents (cf. theorem 5).

8
Boolean algebras of the *n*-gonal theory I

1
THE BOOLEAN ALGEBRAS L_1–L_5

Assuming the data of §1.1, let us form as in §6.2 the polynomial ring $K[x]$ over the given field K, which is also an algebra over K. As before, let k be the number of prime factors of the polynomial $x^n - 1$ in $K[x]$. By chapter 6, theorem 3, $x^n - 1$ is square-free. Therefore $L(x^n - 1)$, the lattice of divisors of $x^n - 1$, is a Boolean algebra with 2^k elements.

Substitution of the cyclic mapping ζ in the polynomials $f(x) \in K[x]$,

$$x // \zeta : f(x) \to f(\zeta), \tag{1}$$

is a homomorphism of $K[x]$ onto $K[\zeta]$, the algebra of cyclic mappings. The kernel of this homomorphism is the ideal $(x^n - 1)$. ($\zeta^n - 1 = 0$, and if $t(x) \in K[x]$ is a proper divisor of $x^n - 1$, then by theorem 4 of chapter 2 $t(\zeta) \neq 0$.) By the homomorphism theorem

$$K[x]/(x^n - 1) \cong K[\zeta] \tag{2}$$

with respect to the isomorphism

$$f(x) + (x^n - 1) \to f(\zeta) \tag{3}$$

by which the residue class of x is replaced by ζ. (The homomorphism (1) is the product of the canonical homomorphism $f(x) \to f(x) + (x^n - 1)$ and the isomorphism (3).)

THEOREM 1 $K[\zeta]$, *the algebra of cyclic mappings, is isomorphic to the residue class ring of the principal ideal domain $K[x]$ with respect to the square-free ideal $(x^n - 1)$.*

If

$$x^n - 1 = p_1(x)p_2(x)...p_k(x) \tag{4}$$

is the prime factorization of $x^n - 1$ in $K[x]$, then by the last expression of the Chinese remainder theorem (chapter 7, theorem 4'', corollary)

$$K[x]/(x^n-1) \cong \sum \oplus K[x]/(p_i(x)) \tag{5}$$

where the terms on the right-hand side are fields.

THEOREM 1′ $K[\zeta]$ *is a direct sum of* k *fields.*

Some important consequences of this theorem are: $E(K[\zeta])$, the Boolean algebra of cyclic projections, has exactly 2^k elements; every ideal in $K[\zeta]$ is a principal ideal generated by a cyclic projection (§7.3). For a more systematic discussion, let us consider the lattices from §7.5 for the principal ideal domain $K[x]$ and the square-free element x^n-1:

L_1: *the lattice of divisors of* x^n-1,
L_2: *the interval* $[(x^n-1), K[x]]$ *of the lattice of ideals of* $K[x]$,
L_3: *the lattice of ideals of* $K[\zeta]$,
L_4: $E(K[\zeta])$, $\circ$, $\cdot$, *the Boolean algebra of cyclic projections,*
L_5: $E(K[\zeta])$, $\circ$, $\cdot$, *the Boolean algebra of cyclic projections.*

In contrast to §7.5, the residue class ring $K[x]/(x^n-1)$ has been replaced by the canonically isomorphic ring $K[\zeta]$ in L_3 and L_4. L_5 is simply a second copy of L_4. L_1–L_4 are Boolean algebras with 2^k elements; there exist natural anti-isomorphisms and isomorphisms among them (chapter 7, theorems 5 and 6) which allow us to determine for any element in one of the Boolean algebras a corresponding element in any one of the others.

In order to describe these mappings let us use the following notations: $t(x)$ denotes a divisor of x^n-1, $\hat{t}(x)$ the complementary divisor determined by $t(x)\hat{t}(x) = x^n-1$, and $e(x)$ a modulo (x^n-1) idempotent polynomial. Elements $e(\zeta)$ are then the cyclic projections. A solution of the congruences

$$e(x) \equiv 1 \quad \mathrm{mod}(t(x)), \tag{6}$$

$$e(x) \equiv 0 \quad \mathrm{mod}(\hat{t}(x)) \tag{7}$$

is modulo (x^n-1) idempotent (see chapter 7, theorem 4′). It is uniquely determined modulo (x^n-1): there is exactly one polynomial of degree $<n$ satisfying the congruences.[1] A solution of the congruences is again denoted by $e_t(x)$.

In order to avoid anti-isomorphisms let us replace the Boolean algebras L_2, L_3, L_4 by their duals, denoted L_2', L_3', L_4'. Thus we are considering the Boolean algebras

$$L_1, L_2', L_3', L_4', L_5 \tag{*}$$

and the following mappings:

i_{12}: $t(x) \to (t(x))$, an anti-isomorphism of L_1 onto L_2, and so an isomorphism of L_1 onto L_2';

---

1 Among the polynomials of degree $<n$ we include the zero polynomial.

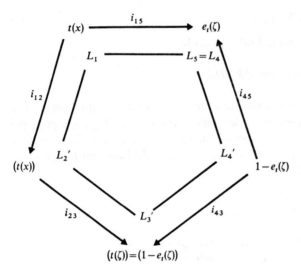

Figure 59

i_{23}: $(t(x)) \to (t(\zeta))$, the canonical isomorphism of L_2 onto L_3, and so of L_2' onto L_3', effected by the substitution $x//\zeta$;

i_{43}: $e(\zeta) \to (e(\zeta))$, the isomorphism of L_4 onto L_3, and so of L_4' onto L_3', originating from chapter 7, theorem 2 (idempotent transfer);

i_{45}: $e(\zeta) \to 1 - e(\zeta)$, the complement formation in $E(K[\zeta])$ (an involutory anti-automorphism), an isomorphism of L_4' onto L_5, and also an isomorphism i_{54} of L_5 onto L_4';

i_{15}: $t(x) \to e_t(\zeta)$, an isomorphism of L_1 onto L_5 (chapter 7, theorem 6).

(The notations L_1, L_2, ..., L_5, i_{12}, i_{23}, i_{43} will in the future be used as defined here, instead of in the more general sense of chapter 7.)

By chapter 7, theorems 5' and 6, we have (see figure 59)

THEOREM 2 $L_1, L_2', L_3', L_4', L_5$ are Boolean algebras with 2^k elements; they are connected by the above isomorphisms.

The product of i_{12}, i_{23} is the mapping $t(x) \to (t(x)) \to (t(\zeta))$ of L_1 onto L_3', and the product of i_{15}, i_{54}, i_{43} is the mapping $t(x) \to e_t(\zeta) \to 1 - e_t(\zeta) \to (1 - e_t(\zeta))$ of L_1 onto L_3'. By the rule of §7.5

$$(t(\zeta)) = (1 - e_t(\zeta)). \tag{8}$$

Therefore $i_{12}, i_{23}, i_{43}{}^{-1}, i_{45}, i_{15}{}^{-1}$ yield a cycle of isomorphisms: their product is the identity on L_1.

Only for the last mapping of theorem 2 is the determination of the image element of a given element non-trivial. Here, for a given divisor $t(x)$ of $x^n - 1$, we must determine a polynomial $e_t(x)$. This can be done explicitly with the help of the derivative $t'(x)$ of $t(x)$:

THEOREM 3 (Derivative formula for $e_t(x)$)   *Let $t(x)$ be a divisor of $x^n - 1$. Then*

$$(1/n)\bar{t}(x)xt'(x) \tag{9}$$

is a polynomial satisfying the congruences (6), (7); *and*

$$(1/n)[\bar{t}(x)xt'(x) - \deg t \cdot (x^n - 1)] \tag{10}$$

is the polynomial of degree $< n$ with this property.

PROOF   Polynomial (9) satisfies congruence (7). It also satisfies congruence (6): for if we differentiate both sides of the equation $\bar{t}(x)t(x) = x^n - 1$ and multiply the resulting equation by x/n, we obtain

$$(1/n)\bar{t}'(x)xt(x) + (1/n)\bar{t}(x)xt'(x) = x^n.$$

The first term $\equiv 0$ modulo $(t(x))$ and the right-hand side $\equiv 1$ modulo $(t(x))$ (since $t(x)|x^n - 1$, $x^n \equiv 1 \bmod(t(x))$).

The term with the highest degree in polynomial (9) is $(1/n) \deg t \cdot x^n$. Therefore (10) is a polynomial of degree $< n$ which is congruent to (9) modulo $(x^n - 1)$.

By theorem 3

$$e_t(\zeta) = (1/n)\bar{t}(\zeta)\zeta t'(\zeta). \tag{11}$$

EXAMPLE   For the divisors $t(x)$: 1, $x - 1$, $x^{n-1} + \ldots + x + 1$, $x^n - 1$ we have the cyclic projections $e_t(\zeta)$: 0, σ, $1 - \sigma$, 1.

To find for a given cyclic mapping $f(\zeta)$ the cyclic projection which generates the same ideal of $K[\zeta]$, we must first determine (say by the Euclidean algorithm) a g.c.d. $t(x)$ of $f(x)$, $x^n - 1$ in $K[x]$. Then by the lemma from §7.5, $(f(\zeta)) = (t(\zeta))$, and by (8) and chapter 6, theorem 5, we have

THEOREM 4   *If $f(\zeta)$ is a cyclic mapping and $t(x)$ the g.c.d. of $f(x)$, $x^n - 1$ in $K[\zeta]$, then*

$$(f(\zeta)) = (t(\zeta)) = (1 - e_t(\zeta)); \tag{12}$$

$1 - e_t(\zeta)$ *is the cyclic projection having the same kernel and image as $f(\zeta)$ and $t(\zeta)$.*

Since i_{15} maps the divisor $\bar{t}(x)$, which is complementary to $t(x)$, onto the cyclic projection $1 - e_t(\zeta)$, which is complementary to $e_t(\zeta)$, $1 - e_t(\zeta) = e_{\bar{t}}(\zeta)$, and hence by (11)

$$1 - e_t(\zeta) = (1/n)t(\zeta)\zeta \bar{t}'(\zeta). \tag{13}$$

Exercise

For $n = 4$ and $n = 6$ determine the polynomials of degree $< n$ in $Q[x]$ which are modulo $(x^n - 1)$ idempotent.

2
DIVISORS OF $x^n - 1$ AND CYCLIC CLASSES

As we know, we can go from the Boolean algebra of cyclic projections to the cyclic classes by Im-transfer: by the proof of the main theorem (§6.2), Im is an isomorphism of L_5: $E(K[\zeta])$, $\circ$, $\cdot$ onto

L_6: *the Boolean algebra of cyclic n-gonal classes.*

However we can also go from the divisors of $x^n - 1$ to the cyclic classes. If $t(x)$ is a divisor of $x^n - 1$, then Ker $t(\zeta)$, the kernel of the cyclic mapping $t(\zeta)$, is a cyclic class by chapter 2, theorem 1. We call it *the cyclic class defined by $t(x)$.* More precisely, it is defined by the cyclic system of equations whose coefficient n-tuple is the coefficient n-tuple of $t(x)$ (cf. §6.3).

THEOREM 5   *The mapping*

$$t(x) \rightarrow \text{Ker } t(\zeta) \tag{14}$$

defines an isomorphism of the lattice of divisors of $x^n - 1$ onto the Boolean algebra of cyclic n-gonal classes.

Our proof of Theorem 5 will use products of previously known isomorphisms. (See figure 60.)

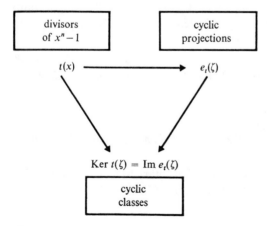

Figure 60

The cyclic projection which projects the set of all n-gons onto the cyclic class defined by $t(x)$ is $e_t(\zeta)$:

$$\text{Ker } t(\zeta) = \text{Im } e_t(\zeta). \tag{15}$$

For $t(\zeta)$ and the cyclic projection $1 - e_t(\zeta)$ have the same kernel by theorem 4; therefore Ker $t(\zeta) = \text{Ker}(1 - e_t(\zeta)) = \text{Im } e_t(\zeta)$ (chapter 2,(7); cf. chapter 6, (12): $e_t(\zeta)$ is, in the notation used there, the cyclic projection $\varepsilon_{t(\zeta)}$).

PROOF OF THEOREM 5　The product of i_{15} and Im, the mapping $t(x) \rightarrow \text{Im } e_t(\zeta)$, defines an isomorphism of $L_1 = L(x^n - 1)$ onto L_6; because of (15) it coincides with (14).

If we think of a system of 2^k non-associated divisors of $x^n - 1$, then from theorem 5 we have

THEOREM 6　*Each of the 2^k cyclic n-gonal classes may be described by a cyclic system of equations whose coefficient n-tuple is the coefficient n-tuple of one of the 2^k divisors of $x^n - 1$.* (Cf. chapter 6, theorem 8.)

THEOREM 7　*Two cyclic systems of equations with coefficient n-tuples $(c_0, c_1, \ldots, c_{n-1})$, $(d_0, d_1, \ldots, d_{n-1})$ define the same cyclic class if and only if the polynomials $\sum c_i x^i$ and $\sum d_i x^i$ in $K[x]$ have the same g.c.d. with $x^n - 1$.*

Thus the question as to whether two cyclic systems of equations define the same class can be answered with the help of the Euclidean algorithm.

PROOF　Ker $f(\zeta) = \text{Ker } g(\zeta)$ is equivalent to $(f(\zeta)) = (g(\zeta))$ by chapter 6, theorem 5, and hence, by the lemma in §7.5, equivalent to the fact that the polynomials $f(x)$ and $g(x)$ in $K[x]$ have the same g.c.d. with $x^n - 1$.

The isomorphisms of theorem 2 give us an isomorphism for each of the Boolean algebras L_1, L_2', L_3', L_4' onto L_5; continuing with the isomorphism Im from L_5 onto L_6, we obtain isomorphisms onto L_6. Hence for each of the Boolean algebras

$$L_1, L_2', L_3', L_4', L_5, \tag{*}$$

we obtain an isomorphism onto L_6, the Boolean algebra of cyclic classes. By these isomorphisms, elements from the Boolean algebras (*) which are related to each other by means of theorem 2 are mapped onto the same cyclic class.

It is of interest to note that each of these isomorphisms of one of the Boolean algebras (*) onto L_6 has a direct meaning, and so gives an approach to the study of cyclic classes. Theorems 5 and 6 show that the isomorphism of L_1 onto L_6 has this property. This and the isomorphism Im of L_5 onto L_6 are the two approaches to the Boolean algebra of cyclic classes which we use the most. The isomorphism of L_4' onto L_6 is the mapping Ker (to each $e(\zeta) \in L_4'$, we associate the cyclic class $\mathrm{Im}(1 - e(\zeta)) = \mathrm{Ker}\, e(\zeta)$, using the product of the isomorphism $1-$ of L_4' onto L_5 and the isomorphism Im of L_5 onto L_6), and therefore has a direct meaning (cf. §5.3). The isomorphisms of L_2' and of L_3' onto L_6 map an ideal onto a cyclic class and will be considered in chapter 9.

Exercise

If $x^n - 1 = t(x)\bar{t}(x)$, then $\mathrm{Im}\, t(\zeta) = \mathrm{Ker}\, \bar{t}(\zeta)$. $t(x) \to \mathrm{Im}\, t(\zeta)$ defines an anti-isomorphism of the lattice of divisors of $x^n - 1$ onto the Boolean algebra of cyclic classes.

3

SPECTRUM

Let N be a splitting field of $x^n - 1$ over K, and w a primitive nth root of unity; thus $N = K(w)$. By the *spectrum of* $x^n - 1$ we shall mean the set of nth roots of unity

$$w, w^2, \ldots, w^{n-1}, w^n = 1. \tag{16}$$

These fall into k classes of conjugates, each class consisting of the zeros of a prime factor of $x^n - 1$. Such a class we call a *spot* on the spectrum.

A function which has (16) as its domain and which gives to each w^l the value 0 or 1 of N, so that conjugate roots of unity have the same value, is called a *characteristic function* of the spectrum. If e, f are characteristic functions, then: ef is equal to 1 on the spots where e and f are both 1, and otherwise it is 0; $e \circ f$ is equal to 1 on the spots where e or f equals 1, and otherwise it is 0. The characteristic functions form a Boolean algebra with 2^k elements, with $\circ$, $\cdot$ as binary operations. The atoms are those functions which are equal to 1 on exactly one spot and 0 everywhere else.

The concept of modulo $(x^n - 1)$ idempotent polynomials of $K[x]$ is intimately connected with the concept of characteristic functions. Congruence modulo $(x^n - 1)$ means equality on the spectrum:

(i) $f(x) \equiv g(x) \bmod(x^n - 1)$ *is equivalent to* $f(w^l) = g(w^l)$ *for all* w^l.

PROOF $h(x) \equiv 0 \bmod(x^n - 1)$, i.e. $x^n - 1 \mid h(x)$ is equivalent to $x - w^l \mid h(x)$ for all w^l, and so equivalent to $h(w^l) = 0$ for all w^l.

From this we have: a polynomial of $K[x]$ is modulo $(x^n - 1)$ idempotent if and only if it has only the values 0 and 1 on the spectrum.

(ii) $e^2(x) \equiv e(x) \bmod(x^n - 1)$ *is equivalent to* $e(w^l) \in \{0, 1\}$ *for all* w^l.

PROOF By (i) the given congruence is equivalent to $e^2(w^l) = e(w^l)$ for all w^l.

Since the $e(w^i)$ are elements of the field N, this is equivalent to saying that all the $e(w^i)$ are 0 or 1.

If a polynomial $f(x) \in K[x]$ has a root of unity as a zero, then all its conjugates are zeros of $f(x)$; if $f(x)$ equals 1 for a root of unity, then the same holds for the conjugates (apply the first statement to $1-f(x)$). Thus from (ii): *The modulo* (x^n-1) *idempotent polynomials of $K[x]$ may be considered as characteristic functions on the spectrum of* x^n-1. Polynomials which are modulo (x^n-1) incongruent are distinct on the spectrum, by (i).

Accordingly by considering the spectrum we obtain a further interpretation for idempotents of the residue class ring $K[x]/(x^n-1)$, and so for the cyclic projections. The construction of a cyclic projection can be traced back to the solution of an interpolation question in $K[x]$. For later use we formulate the following for polynomials over K:

THEOREM 8   *Let $t(x)$ be a divisor of x^n-1. (a) If*

$$e(x) = \begin{cases} 1 \text{ on the zeros of } t(x), \\ 0 \text{ on the remaining } n\text{th roots of unity,} \end{cases}$$

then $\operatorname{Ker} t(\zeta) = \operatorname{Im} e(\zeta)$; *that is, $e(\zeta)$ is the cyclic projection which projects the set of all n-gons onto the cyclic class defined by $t(x)$. (b) If*

$$e(x) = \begin{cases} 0 \text{ on the zeros of } t(x), \\ 1 \text{ on the remaining } n\text{th roots of unity,} \end{cases}$$

then $\operatorname{Ker} t(\zeta) = \operatorname{Ker} e(\zeta)$.

PROOF OF (a)   The conditions for $e(x)$ are equivalent to the congruences (6) and (7). $e(x)$ is thus a polynomial $e_t(x)$. The assertion then follows from (15).

REMARK   The polynomial (10) of theorem 3 solves the interpolation problem of finding a polynomial of degree $<n$ for a divisor of x^n-1 which is equal to 1 on the zeros of the divisor and 0 on the remaining nth roots of unity.

PROOF OF (b)   The conditions for $e(x)$ are equivalent to requiring that $1-e(x)$ satisfy congruences (6) and (7). $1-e(x)$ is thus a polynomial $e_t(x)$. Therefore $e(\zeta) = 1-e_t(\zeta)$ and the assertion follows from theorem 4.

Exercise

A cyclic mapping $f(\zeta)$ is invertible if and only if $f(x)$, x^n-1 are relatively prime in $K[x]$, thus if $f(x)$ (in a splitting field of x^n-1) does not have an nth root of unity as a zero.

4

EXAMPLES OF DEFINING CYCLIC CLASSES BY DIVISORS OF $x^n - 1$

The prime divisor $x-1$ of x^n-1 defines the class $\mathfrak{A}_{1,n}$ of trivial n-gons; for, given an n-gon A, $A \in \mathrm{Ker}(\zeta - 1)$, i.e. $\zeta A = A$, means that all the vertices of A coincide. The divisor of $x^n - 1$ complementary to $x-1$ is

$$m_1(x): = (1/n)(1 + x + x^2 + \dots + x^{n-1})$$

when normalized to have coefficient sum 1; it defines the zero isobaric class $\mathfrak{A}_n$; for $m_1(\zeta) = \sigma$, and $A \in \mathrm{Ker}\,\sigma$ means that $\sigma A = O$. The zero isobaric class is the cyclic class complementary to the class of trivial n-gons.

We can distinguish between two types of divisors $t(x)$ of $x^n - 1$:

I. $x-1 \mid t(x)$, or the equivalent $t(1) = 0$;
II. $x-1 \nmid t(x)$, or the equivalents $t(x) \mid m_1(x)$ and $t(1) \neq 0$.

$t(1)$ is the coefficient sum of $t(x)$.

Each divisor of type I *defines a free cyclic class*; *each divisor of type* II *defines a zero-point class*. (By theorem 5, $x-1 \mid t(x)$ is equivalent to $\mathfrak{A}_{1,n} = \mathrm{Ker}(\zeta - 1) \subseteq \mathrm{Ker}\,t(\zeta)$.) If $t(x)$ is of type I and defines the free cyclic class $\mathfrak{C}$, then $t(x)/(x-1)$ defines the zero-point class $\mathfrak{C}$. ($t(x)/(x-1)$ is the g.c.d. of $t(x)$, $m_1(x)$; it defines $\mathfrak{C} \cap \mathfrak{A}_n = \mathfrak{C}$.)

A divisor $t(x)$ of type II can be normalized[2] by multiplying with $1/t(1)$ so that its coefficient sum becomes 1. The cyclic mapping arising from substituting ζ is then isobaric. A divisor $t(x)$ of type I can be normalized so that $t(x)/(x-1)$ keeps coefficient sum 1.

Let $d \mid n$ and $n = d\bar{d}$. $x^d - 1$ is a divisor of $x^n - 1$ of type I; it defines the class $\mathfrak{A}_{d,\bar{d}}$ of d-gons counted $\bar{d}$ times (since for an n-gon A, $A \in \mathrm{Ker}(\zeta^d - 1)$, i.e. $\zeta^d A = A$ means that A is a d-gon counted $\bar{d}$ times).

$$k_d(x): = (1/d)(1 + x + x^2 + \dots + x^{d-1}) = (1/d)\frac{x^d - 1}{x - 1}$$

is the divisor of $x^n - 1$ (of type II), normalized to have coefficient sum 1, which defines the zero-point class $\mathfrak{A}_{d,\bar{d}}$.

$$m_d(x): = (d/n)(1 + x^d + x^{2d} + \dots + x^{n-d}) = (d/n)\frac{x^n - 1}{x^d - 1}$$

is the divisor of $x^n - 1$ (of type II) complementary to $x^d - 1$, normalized to have coefficient sum 1. It defines the cyclic class complementary to $\mathfrak{A}_{d,\bar{d}}$. By chapter 4, theorem 3, this is the class $\mathfrak{A}_n^d$ of n-gons which are d times isobarically split, and have centre of gravity o. Moreover

2 A modulo $(x^n - 1)$ idempotent divisor of type II is necessarily normalized in this way.
  For since it is $\neq 0$ at spot 1 on the spectrum, it must be equal to 1 there.

$$(x-1)k_d(x)m_d(x) = (1/d)(x^d-1)m_d(x) = (x-1)m_1(x) = (1/n)(x^n-1);$$

the polynomials which appear have been normalized in the above sense.

$m_d(x)$ is equal to 1 on the dth roots of unity and equal to 0 on the remaining nth roots of unity; it is thus modulo (x^n-1) idempotent.
For, if w^l is a dth root of unity so that $w^{ld} = 1$, then

$$m_d(w^l) = (d/n)(1+w^{ld}+w^{2ld}+...+w^{l(n-d)}) = 1;$$

and if w^l is not a dth root of unity so that $w^{ld} \neq 1$, then

$$m_d(w^l) = (d/n)\frac{1-1}{w^{ld}-1} = 0.$$

$m_d(\zeta)$ and $k_d(\zeta)$ are, respectively, the omitting averaging projection μ_d and the consecutive averaging mapping κ_d, whose kernels are known from chapter 4, theorems 1 and 4 and the first rule of §3.2.

The partial products of the product $(x-1)k_d(x)m_d(x)$ are normalized representatives of elements of a Boolean subalgebra of the lattice of divisors of x^n-1. If $d \neq 1, n$, then the eight formally distinct partial products are distinct polynomials. The partial products of $k_d(x)m_d(x)$ define the zero-point classes $\{O\}$, $\mathfrak{A}_{d,\bar{d}}, \mathfrak{A}_n^d, \mathfrak{A}_n$, and the same polynomials multiplied by $x-1$ give the corresponding free classes.

The divisors of x^d-1 define the cyclic n-gonal classes which are contained in the class $\mathfrak{A}_{d,\bar{d}}$ of d-gons counted $\bar{d}$ times. (For $t(x) \mid x^d-1$ is equivalent to Ker $t(\zeta) \subseteq \text{Ker}(\zeta^d-1) = \mathfrak{A}_{d,\bar{d}}$.) These classes form, in the Boolean algebra of cyclic n-gonal classes, a sublattice isomorphic to the Boolean algebra of cyclic d-gonal classes. We obtain the n-gons of these n-gonal classes by counting the d-gons of the cyclic d-gonal classes $\bar{d}$ times.

If $\mathfrak{C}_d$ is an arbitrary cyclic d-gonal class, let the set of n-gons

$$(a_1, ..., a_d, a_1, ..., a_d, ..., a_1, ..., a_d) \quad \text{with } (a_1, ..., a_d) \in \mathfrak{C}_d \tag{17}$$

be denoted by $\mathfrak{C}_{d,\bar{d}}$. Then we have the

LEMMA   Let $d \mid n$, $t(x) \mid x^d-1$, and let $\mathfrak{C}_d$ be the cyclic d-gonal class defined by $t(x)$. Then $\mathfrak{C}_{d,\bar{d}}$ is the cyclic n-gonal class defined by $t(x)$.

PROOF   The d-gonal class defined by x^d-1 is the set $\mathfrak{A}_d$ of all d-gons; the n-gonal class defined by x^d-1 is $\mathfrak{A}_{d,\bar{d}}$.
   Now let

$$t(x) = \sum_{i=0}^{d-1} c_i x^i$$

be a proper divisor of x^d-1. Then $\mathfrak{C}_d$ is the solution space of the cyclic system of equations

E

$$c_0 a_1 + \ldots + c_{d-1} a_d = o, \quad c_0 a_2 + \ldots + c_{d-1} a_1 = o, \quad \ldots \quad (d \text{ equations}). \quad (18)$$

The cyclic n-gonal class $\mathfrak{C}_n$ defined by $t(x)$ is contained in the n-gonal class $\mathfrak{A}_{d,\bar{d}}$; it is the solution space of the cyclic system of equations

$$c_0 a_1 + \ldots + c_{d-1} a_d = o, \quad c_0 a_2 + \ldots + c_{d-1} a_{d+1} = o, \quad \ldots \quad (n \text{ equations}). \quad (19)$$

If $(a_1, \ldots, a_d)$ satisfies system (18), then $(a_1, \ldots, a_d, a_1, \ldots, a_d, \ldots, a_1, \ldots, a_d)$ satisfies system (19); thus $\mathfrak{C}_{d,\bar{d}} \subseteq \mathfrak{C}_n$. Conversely, if $(a_1, \ldots, a_n) \in \mathfrak{C}_n$, then since $\mathfrak{C}_n \subseteq \mathfrak{A}_{d,\bar{d}}$, $a_{d+1} = a_1, \ldots$, and we see from (19) that $(a_1, \ldots, a_d)$ satisfies system (18). Thus $\mathfrak{C}_n \subseteq \mathfrak{C}_{d,\bar{d}}$.

THEOREM 9   *Of the atomic cyclic n-gonal classes, only those which are defined by prime divisors of $F_n(x)$ are not subclasses of proper periodic classes. The remaining atomic cyclic n-gonal classes are obtained from atomic cyclic d-gonal classes with $d \parallel n$ by counting the d-gons of these classes n/d times.*

$d \parallel n$ means that d is a proper divisor of n ($d \mid n$ and $d \neq n$).

PROOF   The atomic cyclic n-gonal classes are defined by the prime divisors of $x^n - 1$, thus by the prime divisors of the cyclotomic polynomials $F_d(x)$ with $d \mid n$; different cyclotomic polynomials have different prime divisors. The cyclic n-gonal class $\mathfrak{C}_n$ defined by a prime divisor $p(x)$ of $F_d(x)$ is contained in the periodic class $\mathfrak{A}_{d,\bar{d}}$, because $F_d(x) \mid x^d - 1$. $p(x)$ already defines a cyclic d-gonal class $\mathfrak{C}_d$, and by the lemma, $\mathfrak{C}_n = \mathfrak{C}_{d,\bar{d}}$. These statements say nothing in the case $d = n$, $\bar{d} = 1$, and since the prime divisors of $F_n(x)$ are not divisors of a polynomial $x^d - 1$ with $d \parallel n$, the n-gonal classes defined by them are not contained in proper periodic classes.

If $K = Q$, then for every n there is exactly one 'typical' atomic cyclic n-gonal class, namely that class defined by $F_n(x)$; all other atomic cyclic n-gonal classes arise from the typical cyclic classes of the proper divisors of n by the process of several countings. The typical atomic cyclic hexagonal class consists of affinely regular hexagons with centre of gravity o. The typical atomic classes for arbitrary n will be described in chapter 10.

Exercises

1   A divisor $t(x)$ of $x^n - 1$ is modulo $(x^n - 1)$ idempotent if and only if the complementary divisor divides $1 - t(x)$.

$$1 - m_d(x) = (d/n)((1 - x^d) + (1 - x^{2d}) + \ldots + (1 - x^{n-d})) \quad (d \mid n)$$

is divisible by $x^d - 1$. Therefore, $m_d(x)$ is modulo $(x^n - 1)$ idempotent.

2   Let $d \mid n$ and $m_d{}'(x)$ be the derivative of $m_d(x)$. Then

$$1 - m_d(x) \equiv (1/d)(x^d - 1)x m_d{}'(x) \quad \mathrm{mod}(x^n - 1).$$

(See chapter 4, (6).)

3 Let $n = d\bar{d}$. If a polynomial $e(x)$ from $K[x]$ is modulo $(x^d - 1)$ idempotent, then $e(x^{\bar{d}})$ is modulo $(x^n - 1)$ idempotent. Using this method, prove that $m_d(x)$ is modulo $(x^n - 1)$ idempotent.

4 If $n = d\bar{d}$ and w is a dth root of unity of K, then

$$\frac{d}{n} w \frac{x^n - 1}{x^d - w} = \frac{d}{n}(1 + w^{-1}x^d + (w^{-1}x^d)^2 + \ldots + (w^{-1}x^d)^{\bar{d}-1})$$

is a modulo $(x^n - 1)$ idempotent proper divisor of $x^n - 1$ in $K[x]$. Can there be other polynomials in $K[x]$ with this property?

9
Boolean algebras of the *n*-gonal theory II

1
GALOIS CORRESPONDENCE OF ANNIHILATORS AND KERNELS

First let us recall a well-known result:

LEMMA   *Let M, N be sets and φ a mapping of M into N, ψ a mapping of N into M, such that*

(a) $\psi\varphi x = x$ *for all $x \in M$,*
(b) $\varphi\psi y = y$ *for all $y \in N$.*

Then φ is a one–one mapping of M onto N and ψ is its inverse.

For the proof note that φ is one–one: for $\varphi x_1 = \varphi x_2$ implies $\psi\varphi x_1 = \psi\varphi x_2$, and thus $x_1 = x_2$ from (a). Also every $y \in N$ is, by (b), the φ-image of $\psi y \in M$.

Now let R be a ring with 1 and $\mathfrak{A}$ an R-module (cf. §7.1). We consider the relation 'r annihilates a' defined on $R \times \mathfrak{A}$ as

$$r \cdot a = o.$$

Every R-submodule $\mathfrak{B}$ of $\mathfrak{A}$, considered as an R-module, has a well-defined *annihilator*:

$$\text{annih } \mathfrak{B} = \{r : r\mathfrak{B} = \{o\}\}$$

which is an ideal of R. Conversely, for every ideal S of R the *kernel*

$$\ker S = \{a : Sa = \{o\}\}$$

is an R-submodule of $\mathfrak{A}$.

The following laws hold for ideals S, T of R and R-submodules $\mathfrak{B}$, $\mathfrak{C}$ of $\mathfrak{A}$:

$$S \subseteq T \text{ implies } \ker S \supseteq \ker T, \tag{1}$$

$$\mathfrak{B} \subseteq \mathfrak{C} \text{ implies annih } \mathfrak{B} \supseteq \text{ annih } \mathfrak{C}, \tag{2}$$

$$\text{annih } \ker S \supseteq S, \tag{3}$$

$$\ker \text{ annih } \mathfrak{B} \supseteq \mathfrak{B}, \tag{4}$$

$$\text{annih ker annih } \mathfrak{B} = \text{annih } \mathfrak{B}, \tag{5}$$

$$\ker \text{ annih ker } S = \ker S. \tag{6}$$

PROOF OF (5)　We obtain $\supseteq$ from (3) by substituting annih $\mathfrak{B}$, and $\subseteq$ by applying annih to (4) and using (2).

Consider now the set of ideals of R which are annihilators of R-submodules of $\mathfrak{A}$:

$$L^{\text{annih}}(R) = \{\text{annih } \mathfrak{B} : \mathfrak{B} \text{ is an } R\text{-submodule of } \mathfrak{A}\},$$

and the set of R-submodules of $\mathfrak{A}$ which are kernels of ideals of R:

$$L^{\text{ker}}(\mathfrak{A}) = \{\ker S : S \text{ is an ideal of } R\}.$$

For every ideal S of R, annih ker S is the least annihilator containing S. (By (3), annih ker S is an annihilator containing S; if $S \subseteq T$ holds for any annihilator T, then by (1), (2), (5), annih ker $S \subseteq$ annih ker $T = T$.) If M is a set of ideals of R and $\langle M \rangle$ the ideal generated by their union, then annih ker $\langle M \rangle$ is the least annihilator which contains all ideals of M.

For every R-submodule $\mathfrak{B}$ of $\mathfrak{A}$, ker annih $\mathfrak{B}$ is the least kernel containing $\mathfrak{B}$. If $\mathfrak{M}$ is a set of R-submodules of $\mathfrak{A}$ and $\langle \mathfrak{M} \rangle$ the R-submodule generated by their union, then ker annih $\langle \mathfrak{M} \rangle$ is the least kernel which contains all R-submodules of $\langle \mathfrak{M} \rangle$.

Every intersection of annihilators is an annihilator, every intersection of kernels is a kernel:

$$\bigcap_{\mathfrak{B} \in \mathfrak{M}} \text{annih } \mathfrak{B} = \text{annih } \langle \mathfrak{M} \rangle, \tag{7}$$

$$\bigcap_{S \in M} \ker S = \ker \langle \mathfrak{M} \rangle. \tag{8}$$

Let us prove (8), writing $\bigcap$ instead of $\bigcap_{S \in M}$. First $\supseteq$: since $S \subseteq \langle M \rangle$, by (1) ker $S \supseteq \ker \langle M \rangle$. Thus, $\cap \ker S \supseteq \ker \langle M \rangle$. Next, $\subseteq$: we have $\cap \ker S \subseteq \ker S$, and so by (2), annih $\cap \ker S \supseteq$ annih ker $S \supseteq S$, the latter by (3). Thus annih $\cap \ker S \supseteq \langle M \rangle$. By (1) we then have that ker annih $\cap \ker S \subseteq \ker \langle M \rangle$, and, because of (4), $\cap \ker S$ is contained in the module on the left.

By (7), $L^{\text{annih}}(R)$ is a complete lattice with inclusion as partial ordering and Inf $M = \bigcap_{S \in M} S$ for every subset M of $L^{\text{annih}}(R)$. Sup $M = \text{annih ker } \langle M \rangle$.

By (8), $L^{\text{ker}}(\mathfrak{A})$ is a complete lattice with inclusion as partial ordering and Inf $\mathfrak{M} = \bigcap_{\mathfrak{B} \in \mathfrak{M}} \mathfrak{B}$ for any subset $\mathfrak{M}$ of $L^{\text{ker}}(\mathfrak{A})$. Sup $\mathfrak{M} = \ker \text{ annih } \langle \mathfrak{M} \rangle$.

THEOREM 1　*If R is a ring with 1 and $\mathfrak{A}$ an R-module, then* ker *(restricted to $L^{\text{annih}}(R)$) and* annih *(restricted to $L^{\text{ker}}(\mathfrak{A})$) are inverse anti-isomorphisms of the lattices $L^{\text{annih}}(R)$ and $L^{\text{ker}}(\mathfrak{A})$.*

PROOF   (5) and (6) show that, according to the lemma, the restriction of ker is a one–one mapping of $L^{\mathrm{annih}}(R)$ onto $L^{\mathrm{ker}}(\mathfrak{A})$ and the restriction of annih is the inverse mapping. By (1) and (2) both mappings reverse inclusions and so are anti-isomorphisms.

(For the general Galois correspondence induced by a relation between two arbitrary sets see G. Birkhoff, *Lattice Theory*, New York, 1940, 1948, 1961).

2

IDEAL-TRANSFER

Now let R be a principal ideal domain and $\mathfrak{A}$ again an R-module. Then the following three expressions hold, the first of which is contained in (8):

(i) $\ker(r) \cap \ker(s) = \ker[(r)+(s)]$,
(ii) $\ker(r)+\ker(s) = \ker[(r) \cap (s)]$,
(iii) if $(0) \neq (r) \subseteq (s)$ and $(r) \in L^{\mathrm{annih}}(R)$, then $(s) \in L^{\mathrm{annih}}(R)$.

These say that *the intersection and sum of two kernels are kernels, and every ideal containing an annihilator $\neq (0)$ is an annihilator.*

PROOF OF (ii) $\subseteq$ comes immediately. For $\supseteq$, let

$$(r)+(s) = (t). \tag{9}$$

If $t = 0$ the assertion is trivial. So let $t \neq 0$. By (9) there exist elements r_1, s_1, u, v with $r = r_1 t, s = s_1 t, t = ur+vs$. ($t$ is a common divisor of r, s which is representable as a linear combination of r, s.) It follows that $t = ur_1 t+vs_1 t$, and thus, since $t \neq 0, 1 = ur_1+vs_1$. On the other hand, $rs_1 = r_1 ts_1 = r_1 s \in (r) \cap (s)$.

Now let $a \in \ker[(r) \cap (s)]$. Then $rs_1 a = r_1 sa = o$, and hence $s_1 a \in \ker(r)$, $r_1 a \in \ker(s)$. Thus

$$a = 1 \cdot a = (ur_1+vs_1)a = ur_1 a+vs_1 a \in \ker(s)+\ker(r).$$

PROOF OF (iii)   We can state this theorem as follows: if $(0) \neq (r) \subseteq (s)$ and annih $\ker(r) = (r)$, then annih $\ker(s) = (s)$. For the proof set annih $\ker(s) = (t)$. By (3) we have $(s) \subseteq (t)$, and by (6), $\ker(s) = \ker$ annih $\ker(s) = \ker(t)$. Since $(r) \subseteq (s)$, there exists an r_1 with

$$r = r_1 s. \tag{10}$$

Therefore for each $a \in \ker(r)$ we have $r_1 sa = o$, and thus $r_1 a \in \ker(s) = \ker(t)$, and so $tr_1 a = o$. Thus tr_1 annihilates $\ker(r)$: $tr_1 \in$ annih $\ker(r) = (r)$. Now let $tr_1 = ur = ur_1 s$. Since $r \neq 0$, by (10) $r_1 \neq 0$, and hence $t = us$; thus $(t) \subseteq (s)$. Thus $(s) = (t)$ and annih $\ker(s) = (s)$.

Theorem 1 and (i)–(iii) yield

THEOREM 2 (Ideal-Transfer) *Let R be a principal ideal domain, and $\mathfrak{A}$ an R-module.
Then $L^{\mathrm{annih}}(R)$ is a sublattice of the lattice of ideals of R, $L^{\mathrm{ker}}(\mathfrak{A})$ is a sublattice of
the lattice of R-submodules of $\mathfrak{A}$, and* ker *is an anti-isomorphism of $L^{\mathrm{annih}}(R)$ onto
$L^{\mathrm{ker}}(\mathfrak{A})$.*

If annih $\mathfrak{A} \neq (0)$, *then $L^{\mathrm{annih}}(R)$ is the finite interval* [annih $\mathfrak{A}$, R] *of the lattice of
ideals of R.*

Let annih $\mathfrak{A} \neq (0)$, and set annih $\mathfrak{A} = (m)$. Then the anti-isomorphism of the
lattice of divisors $L(m)$ onto the interval $[(m), R]$ of the lattice of ideals of R (see
§7.4) defined by $t \to (t)$ (where $t \mid m$) is an anti-isomorphism of $L(m)$ onto
$L^{\mathrm{annih}}(R)$, and $t \to (t) \to \mathrm{ker}(t)$ defines an isomorphism of $L(m)$ onto $L^{\mathrm{ker}}(\mathfrak{A})$.

If, in particular, annih $\mathfrak{A}$ is square-free, then $L(m)$ is a finite Boolean algebra
(§7.4), and $L(m)$, $[(m), R]$, $L^{\mathrm{ker}}(\mathfrak{A})$ are finite Boolean algebras with equally many
elements. (If k is the number of prime factors of m, they each have 2^k elements.)

3
SECOND PROOF OF THE MAIN THEOREM: MAIN DIAGRAM

Returning now to the theory of n-gons, assume again the data of §1.1. If $f(\zeta)$ is a
cyclic mapping and A an n-gon, then

$$f(\zeta)A \tag{11}$$

will, as always, denote the n-gon obtained from A by applying the cyclic mapping
$f(\zeta)$. $\mathfrak{A}_n = V^n$, the vector space of n-gons, is a $K[\zeta]$-module with respect to this
'multiplication.'

But for $n \neq 1$, $K[\zeta]$ is not a principal ideal domain. In order to be able to use
ideal-transfer, we start with the principal ideal domain $K[x]$ and define a product

$$f(x) \cdot A := f(\zeta)A \tag{12}$$

for a polynomial $f(x) \in K[x]$ and an n-gon A. With respect to this multiplication,
$\mathfrak{A}_n$ is a $K[x]$-module with

$$\text{annih } \mathfrak{A}_n = (x^n - 1) \neq (0). \tag{13}$$

By chapter 6, theorem 3, $(x^n - 1)$ is square-free. Therefore

L_2: the interval $[(x^n - 1), K[x]]$ of the lattice of ideals of $K[x]$

is a finite Boolean algebra. By theorem 2 ker is an anti-isomorphism of L_2 onto
the lattice $L^{\mathrm{ker}}(\mathfrak{A}_n)$ of the $K[x]$-module $\mathfrak{A}_n$, and this lattice is a finite Boolean
algebra in the lattice of subspaces of the vector space $\mathfrak{A}_n$.

Now for every ideal $(f(x)) = K[x] \cdot f(x)$ of $K[x]$, we have

$$\mathrm{ker}(f(x)) = \mathrm{Ker} f(\zeta). \tag{14}$$

For $\mathrm{ker}(f(x)) = \{A : K[x] \cdot f(x) \cdot A = \{O\}\} = \{A : f(x) \cdot A = O\} = \{A : f(\zeta)A = O\}$

$= \operatorname{Ker} f(\zeta)$. Since by chapter 2, theorem 1, the kernels of cyclic mappings are the cyclic classes, (14) says: *The lattice L^{ker} ($\mathfrak{A}_n$) of the $K[x]$-module $\mathfrak{A}_n$ consists of the cyclic n-gonal classes.*

Thus we have once more the

MAIN THEOREM   *The cyclic n-gonal classes form a finite Boolean algebra, and this is a sublattice of the lattice of subspaces of the vector space $\mathfrak{A}_n$ (cf. §6.2).*

The first proof of the main theorem depends upon idempotents, the construction of the Chinese remainder theorem, and idempotent-transfer (in the special form of Im-transfer). The second proof avoids these tools and depends upon the concept of an R-module and ideal-transfer.

REMARK   Obviously we also have here a proof for theorem 5 of chapter 8, which expresses the connection between divisors $t(x)$ of $x^n - 1$ and cyclic n-gonal classes. This proof makes no use of cyclic projections (idempotent cyclic mappings). For, as at the end of §2, $t(x) \to (t(x)) \to \ker(t(x))$ defines an isomorphism of the lattice of divisors of $x^n - 1$ onto the Boolean algebra $L^{\mathrm{ker}}(\mathfrak{A}_n)$ which consists of the cyclic n-gonal classes; by (14), $\ker(t(x)) = \operatorname{Ker} t(\zeta)$.

With the anti-isomorphism ker we have a natural approach from L_2 to the Boolean algebra of cyclic n-gonal classes. If we consider $\mathfrak{A}_n$ as a $K[\zeta]$-module, then it is clear from definition (12) that $\ker(f(x)) = \ker(f(\zeta))$, and hence, by (14),

$$\ker(f(x)) = \operatorname{Ker} f(\zeta) = \ker(f(\zeta)). \tag{15}$$

From this we conclude that ker is also an anti-isomorphism of the lattice of ideals of $K[\zeta]$ onto the Boolean algebra of cyclic classes. (Cf. chapter 8, proof of theorem 5.)

Thus we have reached a goal first stated in §8.2; and we formulate a concluding theorem about isomorphisms among the Boolean algebras under consideration. Let the mapping of chapter 8, theorem 5, by which x is replaced by ζ and then the kernel formed, be written $\operatorname{Ker} \cdot x // \zeta$.

THEOREM 3 $\operatorname{Ker} \cdot x // \zeta$, ker, ker, Ker, Im *are isomorphisms of the Boolean algebras*

$$L_1, L_2', L_3', L_4', L_5$$

onto L_6, the Boolean algebra of cyclic n-gonal classes. They map corresponding elements of the Boolean algebras (*) onto the same cyclic class (the correspondences being given by the isomorphisms of theorem 2 of chapter 8).

PROOF   If $t(x)$ is a divisor of $x^n - 1$ and $\dot{K}$ is the multiplicative group of the field K so that $\dot{K} \cdot t(x)$ is the class of divisors associated with $t(x)$, then a quintuple of corresponding elements from the Boolean algebras (*) can be written in the form

$\dot{K} \cdot t(x), \ (t(x)), \ (t(\zeta)) = (e_t'(\zeta)), \ e_t'(\zeta), \ e_t(\zeta)$

(by §8.1), where $1 - e_t(\zeta)$ is denoted by $e_t'(\zeta)$. Then

$$\text{Ker } t(\zeta) = \ker(t(x)) = \ker(t(\zeta)) = \ker(e_t'(\zeta)) = \text{Ker } e_t'(\zeta) = \text{Im } e_t(\zeta).$$

The first, second, and fourth equalities hold because of (15), and the fifth by chapter 2, (7). See figure 61.

Exercises

1 If φ is an endomorphism of a K-vector space and $m(x)$ the minimal polynomial of φ, then $K[\varphi] \cong K[x]/(m(x))$.

2 Let char $K \neq 2$. Then $K[\zeta] = K[\kappa_2]$: every cyclic mapping may be written as a linear combination of powers of κ_2. Give examples. Which is the minimal polynomial of κ_2?

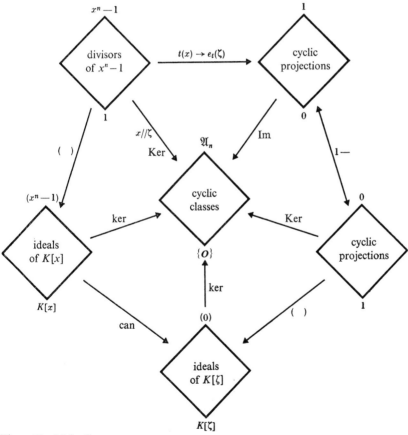

Figure 61   Main diagram

4

GRADUATION: DEGREE OF FREEDOM OF A CYCLIC CLASS

With each of the Boolean algebras $L_1, L_3, L_4 = L_5, L_6$ we can associate one of the numbers of 0, 1, 2, ..., n as degree, dimension, or rank. The natural method for each of these lattices is to use:

L_1: $t(x) \to \deg t$ for divisors of $x^n - 1$, and also the classes of associated divisors (polynomial degree).

L_3: $(f(\zeta)) \to \dim(f(\zeta))$ for ideals of $K[\zeta]$ (the algebra $K[\zeta]$ is an n-dimensional vector space over K by theorem 4 of chapter 2; each ideal is a subspace and so possesses a dimension).

L_5: $e(\zeta) \to \operatorname{rank} e(\zeta)$ for cyclic projections (by the *rank of a cyclic mapping* $f(\zeta) = \sum c_i \zeta^i$ we mean the rank of the cyclic matrix $M(c_0, c_1, ..., c_{n-1})$).

L_6: $\mathfrak{C} \to \deg \mathfrak{C}$ for cyclic classes (degree of freedom of $\mathfrak{C}$; see §1.6).

In this connection we have

$$\operatorname{rank} f(\zeta) = \dim(f(\zeta)), \tag{16}$$

$$n - \deg t = \dim(t(\zeta)) \quad \text{for } t(x) \mid x^n - 1. \tag{17}$$

Let us postpone the proof of these equations and mention first a few consequences. If $t(x)$ is a divisor of $x^n - 1$ and $(f(\zeta)) = (t(\zeta))$, then

$$\deg \operatorname{Ker} f(\zeta) = n - \operatorname{rank} f(\zeta) = n - \dim(f(\zeta)) = \deg t. \tag{18}$$

The first equality holds by chapter 1, theorem 2, and the second and third by (16) and (17).

We now replace the lattice L_3 of ideals of $K[\zeta]$ by its dual lattice L_3' and associate with the ideals their codimensions:

L_3': $(f(\zeta)) \to \operatorname{codim}(f(\zeta))$: $= n - \dim(f(\zeta))$.

Then the 'graduations' of L_1, L_3', L_5, L_6 are the same.

THEOREM 4 *The same number is associated with corresponding elements from the Boolean algebras L_1, L_3', L_5, L_6.*

By 'corresponding elements' we mean elements associated with each other under the isomorphisms of chapter 8, theorem 2, or chapter 9, theorem 3.

PROOF A quadruple of corresponding elements from L_1, L_3', L_5, L_6 can be written in the form

$$\dot{K} \cdot t(x), \quad (t(\zeta)), \quad e_t(\zeta), \quad \operatorname{Ker} t(\zeta) \qquad \text{with } t(x) \mid x^n - 1$$

(chapter 8, theorems 2 and 5). The associated numbers are

$$\deg t, \quad n - \dim(t(\zeta)), \quad \operatorname{rank} e_t(\zeta), \quad \deg \operatorname{Ker} t(\zeta).$$

By (18) the first, second, and fourth numbers are equal. From chapter 8, (8) we find, on replacing $t(x)$ by the complementary divisor $\bar{t}(x)$ (see §8.1, end), that $(e_t(\zeta)) = (\bar{t}(\zeta))$. Therefore by (16) and (17)

$$\operatorname{rank} e_t(\zeta) = \dim(e_t(\zeta)) = \dim(\bar{t}(\zeta)) = n - \deg \bar{t} = \deg t.$$

Let us stress the following statements already contained in theorem 4:

THEOREM 5   *The degree of freedom of the cyclic class defined by a divisor $t(x)$ of $x^n - 1$ is the degree of the polynomial $t(x)$. The degree of freedom of a cyclic class defined by a cyclic system of equations with coefficients $c_0, c_1, \dots, c_{n-1}$ is equal to the polynomial degree of the g.c.d. of $\sum c_i x^i$ and $x^n - 1$ formed in $K[x]$.*

The sum of the degrees of freedom of the atomic cyclic n-gonal classes is always equal to n. This follows from theorem 4 and the fact that the sum of the degrees of the prime factors of $x^n - 1$ is equal to n; it also comes immediately from

THEOREM 6   *The degrees of freedom of cyclic classes $\mathfrak{B}, \mathfrak{C}$ satisfy the relation*

$$\deg(\mathfrak{B} + \mathfrak{C}) + \deg(\mathfrak{B} \cap \mathfrak{C}) = \deg \mathfrak{B} + \deg \mathfrak{C}.$$

PROOF   Since the dimension formula holds for subspaces of the vector space $K[\zeta]$, the assertion is proved using the isomorphism $(f(\zeta)) \to \ker(f(\zeta)) = \operatorname{Ker} f(\zeta)$ of the lattice L_3' onto the Boolean algebra L_6, because $\deg \operatorname{Ker} f(\zeta) = n - \dim(f(\zeta))$.

Referring to §1.8, where a few remarks were made concerning the differences of degree of neighbouring classes in the diagram of hexagonal classes, we can now recognize these differences as a special case of a general phenomenon. We leave the formulation of such a theorem concerning the Boolean algebra of cyclic n-gonal classes to the reader.

PROOF OF EQUATIONS (16) AND (17)   Let us consider the residue class ring $K[x]/(x^n - 1)$ which is canonically isomorphic to $K[\zeta]$. Its elements are residue classes $f(x) + (x^n - 1)$, which we shall abbreviate to $[f(x)]$ $(f(x) \in K[x])$. The special residue classes $[c]$, with $c \in K$, form a subring isomorphic to K. The residue class ring is an algebra over the field of residue classes $[c]$ with $c \in K$, and using the definition

$$c[f(x)] := [cf(x)] \quad \text{for } c \in K,$$

we can interpret it as an algebra over K. Each element of the algebra is uniquely representable in the form $[f(x)]$ with

$$f(x) = \sum_{i=0}^{n-1} c_i x^i, \quad c_i \in K.$$

Thus the algebra is a vector space of dimension n over K, with

$$[1], [x], [x^2], ..., [x^{n-1}] \tag{19}$$

as basis. Every ideal is a principal ideal $([f(x)])$ (see chapter 7, theorem 3) and, as a subspace, has a dimension: $\dim([f(x)])$. We prove that

$$\text{rank } M(c_0, c_1, ..., c_{n-1}) = \dim([\textstyle\sum c_i x^i]), \tag{16'}$$

$$n - \deg t = \dim([t(x)]) \quad \text{for } t(x) \mid x^n - 1. \tag{17'}$$

(16) and (17) then follow, using the isomorphism $[f(x)] \to f(\zeta)$ of $K[x]/(x^n - 1)$ onto $K[\zeta]$.

PROOF OF (16')   Set $c_0 + c_1 x + ... + c_{n-1} x^{n-1} = f(x)$. The residue classes

$$[f(x)], [xf(x)], ..., [x^{n-1} f(x)] \tag{20}$$

form a generating system for the subspace $([f(x)])$. Because

$$[f(x)] = \quad c_0[1] + c_1[x] + ... + c_{n-1}[x^{n-1}],$$
$$[xf(x)] = c_{n-1}[1] + c_0[x] + ... + c_{n-2}[x^{n-1}],$$
$$\cdots\cdots$$
$$[x^{n-1} f(x)] = \quad c_1[1] + c_2[x] + ... + c_0[x^{n-1}],$$

the maximal number of linearly independent generators is equal to the rank of the matrix $M(c_0, c_1, ..., c_{n-1})$.

PROOF OF (17')   The assertion is true for $t(x) = x^n - 1$. Suppose now that $\deg t = m < n$. The residue classes

$$[t(x)], [xt(x)], ..., [x^{n-m-1} t(x)] \tag{21}$$

are linearly independent; for if a linear combination of the residue classes (21) with coefficients $a_0, a_1, ..., a_{n-m-1} \in K$ equals $[0]$, then $[(a_0 + a_1 x + ... + a_{n-m-1} x^{n-m-1}) t(x)] = [0]$, and the first factor must be the zero polynomial.

$x^{n-m} t(x)$ is congruent modulo $(x^n - 1)$ to a polynomial $s(x)$ of degree $< n$. Since $t(x) \mid x^n - 1$, this congruence also holds modulo $(t(x))$, hence $t(x) \mid s(x)$. Thus $s(x)$ can be written as $s_1(x) t(x)$ with $\deg s_1 < n - m$. Thus the residue class

$$[x^{n-m} t(x)] = [s_1(x) t(x)]$$

lies in the subspace T generated by the residue classes (21). Thus $[x]T \subseteq T$.

We now know that $[t(x)] \in T \subseteq ([t(x)])$ and $[x]T \subseteq T$. From this, $T = ([t(x)])$. Therefore the $n - m$ residue classes (21) which are a basis for T are also a basis for $([t(x)])$.

Exercises

1 For $f(x) = \sum_{i=o}^{m} c_i x^i$ ($c_m \neq 0$, $m<n$), deg Ker $f(\zeta) \leq$ deg f comes immediately from the cyclic system of equations $c_0 a_1 + c_1 a_2 + ... + c_m a_{m+1} = o$, ...

2 A less computational proof of (17) may be obtained from

$$n - \dim(t(\zeta)) = \dim K[\zeta]/(t(\zeta)) = \dim K[x]/(t(x)) = \deg t$$

for $t(x) \mid x^n - 1$.

3 Let $K = Q$. Which of the numbers 0, 1, 2, ..., n are degrees of freedom of a cyclic class of n-gons? For which n is each of these numbers a degree of freedom of a cyclic class?

5

MISCELLANEOUS EXERCISES

1 Let $\mathfrak{C}$ be a cyclic class of degree m and $t(x) = c_0 + c_1 x + ... + c_m x^m$ be a divisor of $x^n - 1$ which defines the class $\mathfrak{C}$. Using the equations $c_0 a_1 + ... + c_m a_{m+1} = o$, ..., $c_0 a_{n-m} + ... + c_m a_n = o$, we obtain a normal representation of the n-gons (a_1, ..., a_m, a_{m+1}, ..., a_n) of $\mathfrak{C}$ with a_1, ..., a_m as parameters by expressing a_{m+1}, ..., a_n recursively in terms of a_1, ..., a_m. (Here $t(x)$ should be normalized to be monic.) Let γ be an automorphism, φ an endomorphism, of an abelian group. Denoting $\gamma \varphi \gamma^{-1}$ by φ^γ, show that

$$\gamma(\text{Ker } \varphi) = \text{Ker } \varphi^\gamma, \qquad \gamma(\text{Im } \varphi) = \text{Im } \varphi^\gamma.$$

3 Let $n = 2m$. α: (a_1, a_2, a_3, ..., a_{2m}) $\rightarrow$ (a_1, $-a_2$, a_3, ..., $-a_{2m}$) is an involutory automorphism of $\mathfrak{A}_n$ but not a cyclic mapping. For every cyclic mapping, $f(\zeta)^\alpha = f(-\zeta)$, and

$$f(\zeta) \rightarrow f(-\zeta) \tag{*}$$

is an involutory automorphism of $K[\zeta]$ and also of $E(K[\zeta])$. Then $\alpha(\text{Ker } t(\zeta)) = \text{Ker } t(\zeta)^\alpha = \text{Ker } t(-\zeta)$, and

$$\text{Ker } t(\zeta) \rightarrow \text{Ker } t(-\zeta) \tag{**}$$

is an involutory automorphism of the Boolean algebra of cyclic classes which preserves the class degree. (**) interchanges at least the ASO class and the zero isobaric class, and also the class of trivial $2m$-gons and the class of digons (with centre of gravity o) counted m times.

   For $K = Q$, $n = 4$, and $n = 6$, investigate how (*) permutes the cyclic projections and (**) the cyclic classes.

4 Let n be even and $d \mid n$. Since $m_d(x)$ is a modulo ($x^n - 1$) idempotent divisor of $x^n - 1$ (§8.4), the same is true for $m_d(-x)$. For even d, $m_d(x) = m_d(-x)$. Suppose now that d is odd. Produce relations between $m_d(x)$ and $m_d(-x)$ and explain them

in the Boolean algebras of cyclic projections and of cyclic classes. For example:

$$m_d(x) + m_d(-x) = m_{2d}(x), \quad m_d(x)m_d(-x) \equiv 0 \quad \mod(x^n - 1), \tag{*}$$

$$1 - m_d(-x) + m_1(-x) = 1 - (m_{2d}(x) - m_d(x)) + (m_2(x) - m_1(x)). \tag{**}$$

If in (**) we replace x by ζ, then we obtain the cyclic projection which has the same image and kernel as $\alpha_d = 1 - \zeta + \zeta^2 - \ldots + \zeta^{d-1}$; it can also be written as $\mu_2 \circ \mu_d \circ (1 - \mu_{2d})$.

5 Let char $K \neq 2$. If n is odd, α_d $(d \mid n)$ is invertible with inverse $\kappa_2(1 - \zeta^d + \zeta^{2d} - \ldots + \zeta^{n-d})$.

6 $\alpha_3 = 1 - \zeta + \zeta^2$ specializes the set of n-gons only for $n = 6, 12, 18, \ldots$.

7 α_3 maps every parallelogram onto itself, up to a cyclic permutation of the vertices (and so also every parallelogram counted several times). Are there other non-trivial n-gons with this property?

8 Let char $K = 0$. Consider $\kappa_r = (1/r)(1 + \zeta + \zeta^2 + \ldots + \zeta^{r-1})$ not only for divisors r of n, but also for all $r = 1, 2, \ldots$. If d is the g.c.d. of r and n, then κ_r, κ_d have the same image and kernel.

If u is odd and d the g.c.d. of u, n, then α_u, α_d have the same image and kernel.

9 $\zeta \to \zeta^{-1}$ effects an automorphism of $K[\zeta]$ which maps $f(\zeta) = c_0 + c_1\zeta + \ldots + c_{n-1}\zeta^{n-1}$ onto $f(\zeta^{-1}) = c_0 + c_{n-1}\zeta + \ldots + c_1\zeta^{n-1}$. Let $t(x)$ be a divisor of $x^n - 1$. If $t(x)$ is symmetric or anti-symmetric (see §12.1; for example, this condition is always fulfilled for $K = Q$), then $t(\zeta)$ and $t(\zeta^{-1})$ have the same image and kernel.

10 *Cyclic classes of cyclic mappings* By definition (§2.1), a cyclic mapping of the vector space V^n is given by an n-tuple $(c_0, c_1, \ldots, c_{n-1})$ with $c_i \in K$, thus by an 'n-gon' of the vector space K^n. The concepts of the n-gonal theory defined in K^n (the case $V = K$) may then be taken from K^n over to $K[\zeta]$ by the vector space isomorphism $(c_0, c_1, \ldots, c_{n-1}) \to \Sigma c_i\zeta^i$. The ideals of $K[\zeta]$ can then be interpreted as 'cyclic classes of cyclic mappings.'

EXAMPLE The ideal (σ) consists of the 'trivial' cyclic mappings, i.e. the $\Sigma c_i\zeta^i$ with $c_0 = c_1 = \ldots = c_{n-1}$; the ideal $(1-\sigma)$ consists of the cyclic mappings 'with centre of gravity 0,' i.e. the $\Sigma c_i\zeta^i$ with $(1/n)\Sigma c_i = 0$. (Cf. §3.3, exercise.)

11 *Cyclic mappings of a cyclic class* Let $\mathfrak{C}$ be a cyclic class. By chapter 2, theorem 5, all cyclic mappings of $\mathfrak{A}_n$ map the class $\mathfrak{C}$ into itself. But, in general, many of them will map the n-gons of $\mathfrak{C}$ equally. Determine subsets of $K[\zeta]$ which represent all cyclic mappings of $\mathfrak{C}$.

Let $\mathfrak{C} = \text{Ker } t(\zeta) = \text{Im } e_t(\zeta)$, where $t(x)$ is a divisor of $x^n - 1$. Two cyclic mappings map $\mathfrak{C}$ equally if and only if they are congruent modulo $(t(\zeta))$. Different cyclic mappings of $\mathfrak{C}$ can be represented by elements of the ideal $(e_t(\zeta))$; $K[\zeta] = (t(\zeta)) \oplus (e_t(\zeta))$ and $(e_t(\zeta)) \cong K[\zeta]/(t(\zeta)) \cong K[x]/(t(x))$.

$(e_t(\zeta))$ consists of the cyclic mappings which map $\mathfrak{A}_n$ into $\mathfrak{C}$.

12 Let $n = 4$. Different cyclic mappings of the class of parallelograms, i.e. the quadrangles

$$(a_1, a_2, a_3, a_4) \quad \text{with } a_1 - a_2 + a_3 - a_4 = o,$$

are represented by cyclic mappings with coefficient quadruples

(c_0, c_1, c_2, c_3) with $c_0 - c_1 + c_2 - c_3 = 0$,

or, in the sense of exercise 10, the 'parallelograms' of the cyclic mappings. Among these we have, in particular, the cyclic mappings which map the set of all quadrangles onto the set of parallelograms, for example κ_2 and the projection $1 - \mu_2 + \sigma$ with coefficient quadruples $\frac{1}{2}(1, 1, 0, 0)$ and $\frac{1}{4}(3, 1, -1, 1)$.

For $n = 4$ establish a general theorem, originating from this example, concerning distinct cyclic mappings of a cyclic n-gonal class.

13 KINDER   From any permutation π of the set $\{1, ..., n\}$ we obtain an automorphism $\bar{\pi}$ of the K-vector space $\mathfrak{A}_n = V^n$ of n-gons by permuting the n vertices of each n-gon according to π:

$$\bar{\pi}: (a_1, ..., a_n) \to (a_{\pi(1)}, ..., a_{\pi(n)}).$$

$\pi \to \bar{\pi}$ is a monomorphism of the group $\mathfrak{S}_n$ of all permutations of $\{1, ..., n\}$ into the automorphism group of $\mathfrak{A}_n$. For example, the cyclic mapping ζ is the image of the permutation

$$\begin{pmatrix} 1 & 2 & ... & n \\ 2 & 3 & ... & 1 \end{pmatrix} = (1 \ 2 \ ... \ n).$$

In general, $\bar{\pi}$ is not a cyclic mapping. Show that for each $\pi \in \mathfrak{S}_n$ the following conditions are equivalent:

(1) $\bar{\pi} \in K[\zeta]$,
(2) $\bar{\pi}\zeta = \zeta\bar{\pi}$,
(3) $\pi (1...n) = (1...n)\pi$,
(4) $(\pi(1)...\pi(n)) = (1...n)$,
(5) $\pi = (1...n)^j$ for some $j \in \{0, 1, ..., n-1\}$,
(6) $\bar{\pi} = \zeta^j$ for some $j \in \{0, 1, ..., n-1\}$.

Therefore the only cyclic mappings of the form $\bar{\pi}$ are the powers of ζ.

14 KINDER   By exercise 2 an automorphism of $\mathfrak{A}_n$ maps cyclic classes onto cyclic classes if it belongs to the normalizer $N(K[\zeta])$ of $K[\zeta]$ in End $(\mathfrak{A}_n)$. Show that for any permutation π of $\{1, ..., n\}$ the following conditions are equivalent:

(1) $\bar{\pi} \in N(K[\zeta])$,
(2) $\bar{\pi}\zeta\bar{\pi}^{-1} \in K[\zeta]$,
(3) $\overline{(\pi(1)...\pi(n))} \in K[\zeta]$,
(4) $(\pi(1)...\pi(n)) = (1...n)^j$ for some $j \in \{0, 1, ..., n-1\}$;

(5) there exists some j relatively prime to n and some k in $\{0, 1, ..., n-1\}$ such that for $i = 1, 2, ..., n$ the following holds:

$$\pi(i) \equiv i \cdot j + k \quad \mathrm{mod} \ n. \tag{*}$$

Let G_n denote the group of permutations π with (1)–(5). Let F_n be the group of $\pi \in G_n$ for which the automorphism $\bar{\pi}$ maps every cyclic class onto itself. The permutation $(1 \ldots n) \in F_n$ $(j = k = 1)$ generates a group Z_n of order n which is a normal subgroup in G_n and therefore a fortiori in F_n.

Show that the factor group F_n/Z_n is isomorphic to the Galois group of the polynomial $x^n - 1$ over the field K. (The residue class πZ_n of a $\pi \in F_n$ with (*) corresponds, if w is a primitive nth root of unity in a splitting field of $x^n - 1$ over K, to the K-automorphism $\sum c_i w^i \rightarrow \sum c_i w^{i \cdot j}$ $(c_i \in K)$ of $K(w)$.)

15 Let K_p be the prime field of characteristic p and p a divisor of n. If $n = p^l m$ with $(p, m) = 1$, then $x^n - 1 = (x^m - 1)^{p^l}$. There are at least $(p^l + 1)^{\tau(m)}$ cyclic classes. Determine the cyclic hexagonal classes over K_3. They form a distributive lattice but not a Boolean algebra.

PART IV
Atomic decompositions

10
Rational components of an *n*-gon

1
Q-REGULAR *n*-GONS

An *n*-gon A is said to be *Q-regular* if, for each divisor $d \neq 1$ of n, all omitting sub-d-gons of A are isobaric.[1] Let us denote the set of *Q*-regular *n*-gons by $\Re_n$.

For the divisor $d = n$ this definition merely requires A to be isobaric to itself. Since this is always the case, we can limit ourselves to non-trivial divisors d of n ($d \neq 1, n$). Every monogon, and every *p*-gon (prime *p*), is *Q*-regular: $\Re_1 = \mathfrak{A}_1$, $\Re_p = \mathfrak{A}_p$. Thus the concept of *Q*-regularity says nothing for monogons and *p*-gons. In order to see the strength or weakness of our requirement for composite *n*, it is useful to sketch a few *Q*-regular *n*-gons. Figures 62–85 give some suggestions for this.

In chapter 4 we introduced classes of isobarically splitting *n*-gons: if d is a divisor of *n*, then $\mathfrak{A}_n^{n/d}$ is the class of *n*-gons which are n/d times isobarically split, that is, whose omitting sub-*d*-gons are isobaric. In this notation, $\mathfrak{A}_n^1$ is the class of all *n*-gons, $\mathfrak{A}_n^n$ the class of trivial *n*-gons.

$\Re_n$ is the intersection of all these classes up to the class $\mathfrak{A}_n^n$:

$$\Re_n = \bigcap_{\substack{d|n \\ d \neq 1}} \mathfrak{A}_n^{n/d} = \bigcap_{\substack{d|n \\ d \neq n}} \mathfrak{A}_n^d. \tag{1}$$

(In (1), $d \neq 1, n$ could be written both times.) $\Re_n$, as an intersection of free cyclic classes, is a free cyclic class.

EXAMPLE $\Re_4$ is the class $\mathfrak{A}_4^2$ of parallelograms; $\Re_6 = \mathfrak{A}_6^2 \cap \mathfrak{A}_6^3$ is the class of affinely regular hexagons (§4.1). In general, *Q*-regular 2*p*-gons (*p* an odd prime) are the 2*p*-parallelograms with alternating vertex sum o. These are also the 2*p*-gons in which all *p*-tuples of consecutive vertices have the same alternating sum. The 'parametric representation' is

$$(a_1, \dots, a_p, -a_1 + 2s, \dots, -a_p + 2s) \quad \text{with } s = a_1 - a_2 + a_3 - \dots + a_p.$$

The *Q*-regular 2*p*-gons are obtained by choosing *p* points arbitrarily, constructing

---

[1] The omitting sub-*d*-gons of A are the rows of the vertex scheme, modulo n/d, of A.

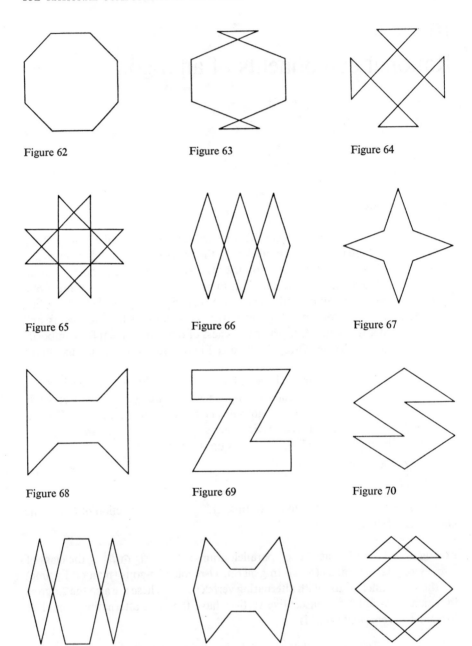

Figure 62

Figure 63

Figure 64

Figure 65

Figure 66

Figure 67

Figure 68

Figure 69

Figure 70

Figure 71

Figure 72

Figure 73

Q-regular 8- and 10-gons

Figure 74

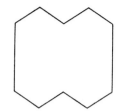

Figure 75

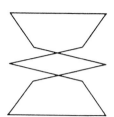

Figure 76

Figure 77

Figure 78

Figure 79

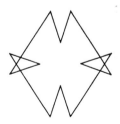

Figure 80

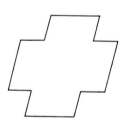

Figure 81

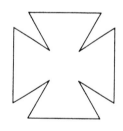

Figure 82

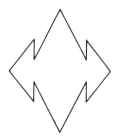

Figure 83

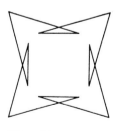

Figure 84

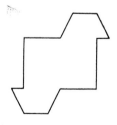

Figure 85

Q-regular 10- and 12-gons

their alternating sum *s* by repeated formation of fourth parallelogram points, and then reflecting the chosen points in *s*.

THEOREM 1   Let $n \neq 1$. An n-gon **A** is **Q**-regular if and only if, for each prime divisor p of n, all omitting sub-p-gons of **A** are isobaric.

PROOF   If $t \mid d \mid n$, then $\mathfrak{A}_n^{n/t} \subseteq \mathfrak{A}_n^{n/d}$. Since for every divisor $d \neq 1$ of *n*, there is a prime divisor *p* of *n* with $p \mid d$, there is, for every class appearing in the first intersection of (1), a class $\mathfrak{A}_n^{n/p}$ contained in it. Therefore

$$\mathfrak{R}_n = \bigcap_{p \mid n} \mathfrak{A}_n^{n/p} \quad \text{for } n = 1. \tag{2}$$

EXAMPLE $\mathfrak{R}_8$ is the class $\mathfrak{A}_8^4$ of 8-parallelograms (see chapter 1, theorem 4).

Omitting sub-*d*-gons of **Q**-regular *n*-gons are **Q**-regular $(d \mid n)$. This depends upon the fact that, if $t \mid d \mid n$, every omitting sub-*t*-gon of an omitting sub-*d*-gon of an *n*-gon **A** is an omitting sub-*t*-gon of **A**. Conversely, we can ask how much can be concluded about the **Q**-regularity of a complete *n*-gon from the **Q**-regularity and isobarism of its omitting subpolygons. For example, the following holds: choose two isobaric affinely regular 6-gons and label them as $(a_1, a_3, ..., a_{11})$, $(a_2, a_4, ..., a_{12})$. Then $(a_1, a_2, ..., a_{12})$ is a **Q**-regular 12-gon. Special choices of affinely regular 6-gons give us the Cross Formée (figure 82) and the Star of David (figure 77).

The **Q**-regular *n*-gons with centre of gravity *o* form a cyclic zero-point class $\mathfrak{R}_n$, which can be represented by equations corresponding to (1) or (2) with the help of classes $\mathfrak{A}_n^{n/d}$ or $\mathfrak{A}_n^d$ respectively.

Exercises

1 If *n* is a power of a prime *p*, then all *n*-gons whose omitting sub-*p*-gons are isobaric are **Q**-regular.
2 If $n \neq 1$ and n^* is the square-free kernel of *n* (the largest square-free divisor of *n*), then the **Q**-regular *n*-gons are precisely those *n*-gons whose omitting sub-n^*-gons are **Q**-regular and isobaric.

2
CYCLIC CLASSES DEFINED BY CYCLOTOMIC POLYNOMIALS

Continuing from chapter 8, we determined the cyclic *n*-gonal classes defined by the cyclotomic polynomials $F_d(x)$ with $d \mid n$, starting with the class defined by $F_n(x)$. For $n = 1$ this is the class of all 1-gons.

THEOREM 2   *Let $n \neq 1$. The cyclic n-gonal class defined by $F_n(x)$ consists of the Q-regular n-gons with centre of gravity o:*

$$\text{Ker } F_n(\zeta) = \mathfrak{R}_n. \tag{3}$$

PROOF   From the polynomials $m_d(x)$ whose behaviour on the spectrum of $x^n - 1$ is known (§8.4), form the Boolean sum

$$\sum_{d \| n} \circ \, m_d(x) \tag{4}$$

according to chapter 5,(1). Polynomial (4) is equal to 0 on the primitive nth roots of unity and 1 on the remaining nth roots of unity. For on the primitive nth roots of unity each term is equal to 0. And on the dth roots of unity with $d \| n$, the term $m_d(x)$ equals 1, and so the Boolean sum is also 1.

    Therefore, by chapter 8, theorem 8b, since $m_d(\zeta) = \mu_d$, the first equality of the following holds:

$$\text{Ker } F_n(\zeta) = \text{Ker } \sum_{d \| n} \circ \, \mu_d = \bigcap_{d \| n} \text{Ker } \mu_d = \bigcap_{d \| n} \mathfrak{A}_n^d = \mathfrak{R}_n.$$

The second equality holds by rule (18) of chapter 5, and the third by §4.3.

THEOREM 3   *$F_1(x)$ defines the class of trivial n-gons; $F_d(x)$, with $d \,|\, n$, $d \neq 1$, defines the class of Q-regular d-gons with centre of gravity o, counted $\bar{d}$ times:*

$$\text{Ker } F_1(\zeta) = \mathfrak{A}_{1,n}, \qquad \text{Ker } F_d(\zeta) = \mathfrak{R}_{d,\bar{d}} \quad \text{for } d \,|\, n, \, d \neq 1. \tag{5}$$

PROOF   The first statement is known from §6.3. Let $d \,|\, n$, $d \neq 1$. Then $F_d(x)$ defines a d-gonal class, and theorem 2 tells us it is the d-gonal class $\mathfrak{R}_d$. Therefore, by the lemma from §8.4, $F_d(x)$ defines the n-gonal class $\mathfrak{R}_{d,\bar{d}}$.

    One of these classes is associated with each divisor d of n; by chapter 9, theorem 5, its degree of freedom is the degree of the defining polynomial $F_d(x)$, and hence the value $\varphi(d)$ of the Euler φ-function. Incidentally, we see that for $n \neq 1$ the (free) class of Q-regular n-gons has degree of freedom $\varphi(n) + 1$.

    Since the cyclotomic polynomials $F_d(x)$ with $d \,|\, n$ have product $x^n - 1$ and are relatively prime, $\mathfrak{A}_n$ is the direct sum of cyclic classes (5) because of chapter 8, theorem 5.

    If $K = Q$, the $F_d(x)$ with $d \,|\, n$ are the monic prime factors of $x^n - 1$, and therefore the classes (5) are the atomic cyclic classes (chapter 8, theorem 5; see §6.3).

THEOREM 4   *For $K = Q$, the atomic cyclic n-gonal classes are: the class $\mathfrak{A}_{1,n}$ of trivial n-gons; and the classes $\mathfrak{R}_{d,\bar{d}}$ of Q-regular d-gons with centre of gravity o ($d \,|\, n$, $d \neq 1$) counted $\bar{d}$ times, among which, for $d = n$, $n \neq 1$, is the class $\mathfrak{R}_n$ of Q-regular n-gons with centre of gravity o.*

EXAMPLE $K = Q, n = 12$. The six atomic cyclic classes are:

the trivial 12-gons,
the 2-gons with centre of gravity o, counted 6 times,
the 3-gons with centre of gravity o, counted 4 times,
the parallelograms with centre of gravity o, counted 3 times,
the affinely regular 6-gons with centre of gravity o, counted twice,
the Q-regular 12-gons with centre of gravity o.

Since $F_{12}(x) = x^4 - x^2 + 1$, the Q-regular 12-gons with centre of gravity o are, by theorem 2, the solutions of the cyclic system of equations $a_1 - a_3 + a_5 = 0, \ldots$, and are thus the 12-gons $(a_1, a_2, \ldots, a_{12})$ whose two omitting sub-6-gons $(a_1, a_3, \ldots, a_{11})$ and $(a_2, a_4, \ldots, a_{12})$ are affinely regular 6-gons with centre of gravity o.

3
RATIONAL COMPONENTS OF AN *n*-GON

For every cyclotomic polynomial $F_d(x)$ with $d \mid n$ there exists a uniquely determined cyclic projection mapping the set of all *n*-gons onto the cyclic class defined by $F_d(x)$ (see §8.2). We now wish to denote this by ε_d, so that

$$\text{Ker } F_d(\zeta) = \text{Im } \varepsilon_d. \qquad (6)$$

ε_d is the image of $F_d(x)$ with respect to the isomorphism i_{15} of the lattice of

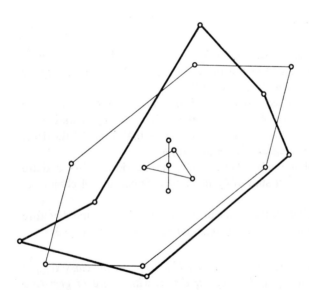

Figure 86   Decomposition of a 6-gon with centre of gravity o into its rational components

divisors of $x^n - 1$ onto the Boolean algebra of cyclic projections, as considered in chapter 8. Since the cyclotomic polynomials $F_d(x)$ with $d \mid n$ have product $x^n - 1$ and are relatively prime, this isomorphism shows that the ε_d's with $d \mid n$ have sum 1 and are mutually orthogonal.

For each n-gon A

$$A = \sum_{d \mid n} \varepsilon_d A \tag{7}$$

is the unique decomposition of A into n-gons from the cyclic classes (5). Here $\varepsilon_1 = \sigma$ (the cyclic projection mapping the set of all n-gons onto the class of trivial n-gons is σ), and thus $\varepsilon_1 A$ is the centre-of-gravity n-gon of A. For $d \neq 1$, $\varepsilon_d A$ is a Q-regular d-gon with centre of gravity o, counted $\bar{d}$ times; in particular, for $n \neq 1$, $\varepsilon_n A$ is a Q-regular n-gon with centre of gravity o.

If $K = Q$, this decomposition of A is the atomic decomposition of A. In general, we call the n-gons $\varepsilon_d A$ the *rational components of A* and the n-gon $\varepsilon_n A$ the *Q-regular component of A*. See figure 86 for $n = 6$.

For every divisor d of n we can get the cyclic projection ε_d from $F_d(x)$ by the derivative formula (11) of chapter 8. But there is another interesting representation of ε_d:

THEOREM 5 $\varepsilon_d = \mu_d \prod_{t \| d} (1 - \mu_t)$.

To prove this, we form for $d \mid n$ the polynomial

$$f_d(x) := m_d(x) \prod_{t \| d} (1 - m_t(x)).$$

It is equal to 1 on the primitive dth roots of unity and 0 on the remaining nth roots of unity. For, on the primitive dth roots of unity all factors of the product $f_d(x)$ equal 1, since on such a root of unity $m_d(x)$ is equal to 1 and $m_t(x)$ equals 0 for all $t \| d$. The remaining nth roots of unity are either tth roots of unity for a $t \| d$ or not dth roots of unity. For the former, $m_t(x)$ equals 1 and so $1 - m_t(x)$ equals 0; for the latter, $m_d(x)$ is 0.

Thus by chapter 8, theorem 8a, $f_d(\zeta)$ is the cyclic projection which projects the set of all n-gons onto the cyclic class defined by $F_d(x)$, and so is equal to ε_d.

As an addition to theorem 5 we have the following inversion formula:

$$\mu_d = \sum_{t \mid d} \varepsilon_t. \tag{8}$$

(Since the ε_d's are mutually orthogonal, the sums in (8) can also be written as a Boolean sum.)

PROOF   The polynomials $m_d(x)$ and $f_d(x)$ are connected by the congruence

$$m_d(x) \equiv \sum_{t \mid d} f_t(x) \quad \mod(x^n - 1). \tag{9}$$

For, since $f_t(x)$ equals 1 on the primitive tth roots of unity and 0 on the remaining nth roots of unity, the sum on the right is equal to 1 on all the dth roots of unity and 0 on all the other nth roots of unity, thus coinciding with $m_d(x)$ on the spectrum of $x^n - 1$. Replacing x by ζ in (9) yields (8).

Exercise

The multiplicative rule for μ_d of §4.2 follows immediately from (8).

4

THE BOOLEAN ALGEBRA GENERATED BY OMITTING AVERAGING PROJECTIONS AND ITS ATOMS

The Boolean algebra $E(K[\zeta])$, $\circ$, $\cdot$ of cyclic projections contains the omitting averaging projections μ_d with $d \mid n$ and the Boolean subalgebra E_μ generated by them (§5.5). But the μ_d's do not generate E_μ freely, i.e. not every Boolean minimal polynomial of μ_d is an atom of E_μ.

THEOREM 6   *The cyclic projections ε_d ($d \mid n$) are the atoms of the Boolean algebra E_μ generated by the omitting averaging projections μ_d ($d \mid n$). E_μ has $2^{\tau(n)}$ elements, and in the case $K = Q$ consists of all cyclic projections.*

PROOF   By theorem 5 the ε_d's ($d \mid n$) lie in E_μ. They have sum 1 and are mutually orthogonal; since they are also $\neq 0$, by chapter 5, theorem 2, they are the atoms of a Boolean subalgebra of E_μ with $2^{\tau(n)}$ elements. Because of (8) this Boolean subalgebra of E_μ is equal to E_μ. In the case $K = Q$, the number of prime factors of $x^n - 1$ is equal to $\tau(n)$ (§6.2), and the number of cyclic projections is equal to $2^{\tau(n)}$ (chapter 8, theorem 1′, theorem 2). This gives the last assertion.

Combinations of the μ_d's can be reduced in many cases by using the multiplicative rule from §4.2. If, as in theorem 5 and also in (4), we write combinations of the μ_d's with the help of operations $\circ$, $\cdot$, $1-$ given in $E(K[\zeta])$, then under the isomorphism Im or anti-isomorphism Ker, they are mapped onto combinations of cyclic classes, which is a distinct advantage. Also, in this notation the product of theorem 5 can in general be reduced, since we can limit ourselves, when $d \neq 1$, to the maximal proper divisors t of d. On the other hand, we can multiply out the product of theorem 5 in $K[\zeta]$ and obtain, by the above rule, ε_d as an integral linear combination of the μ_t's with $t \mid d$.

EXAMPLE $n = 12$: $\varepsilon_1 = \mu_1$, $\varepsilon_2 = \mu_2 - \mu_1$, $\varepsilon_3 = \mu_3 - \mu_1$, $\varepsilon_4 = \mu_4 - \mu_2$,
$\varepsilon_6 = \mu_6 - \mu_3 - \mu_2 + \mu_1$, $\varepsilon_{12} = \mu_{12} - \mu_6 - \mu_4 + \mu_2$ ($\mu_1 = \sigma$ always, and, for $n = 12$, $\mu_{12} = 1$).

In this example, only 1, 0, -1 appear as coefficients of the μ_t's ($t \mid d$) in the representation of ε_d, and for $d \neq 1$ the sum of the coefficients is equal to 0. This holds generally. For, from (8), we obtain the following with the help of the Möbius inversion formula:[2]

THEOREM 7 $\qquad \varepsilon_d = \sum_{t|d} \mu\left(\dfrac{d}{t}\right) \mu_t.$

In this connection $\mu(m)$ is the *Möbius μ-function*, defined for natural numbers m as follows: $\mu(1) = 1$, $\mu(m) = (-1)^r$ if m is the product of r distinct prime factors, $\mu(m) = 0$ if $m \neq 1$ and is not square-free. Then $\sum_{t|m} \mu(t) = 0$ for $m \neq 1$.

Theorems 5 and 7 contain, in particular, a representation of ε_n if we remember that $\mu_n = 1$. We also note a representation of $1 - \varepsilon_n$ yielded by these theorems:

$$1 - \varepsilon_n = \sum_{d \| n} \circ \mu_d, \tag{10}$$

$$1 - \varepsilon_n = -\sum_{d \| n} \mu\left(\frac{n}{d}\right) \mu_d. \tag{11}$$

For (10) we must use one of the de Morgan laws. For (11) we need only use the fact that ε_n contains the term 1 in its representation in theorem 7.

Exercise

The projection ε_d ($d \mid n$) onto the cyclic class defined by $F_d(x)$ may be represented with the help of the Möbius μ-function as follows:

$$\varepsilon_d = \sum_{i=0}^{n-1} c_i \zeta^i \quad \text{with } c_i = \frac{1}{n} \sum_{t|(d,i)} \mu\left(\frac{d}{t}\right) t$$

$((d, i)$ denotes the g.c.d. of d, i). If in particular d is square-free, then it can also be written with Euler's φ-function as

$$c_i = \frac{1}{n} \mu\left(\frac{d}{(d, i)}\right) \varphi((d, i)).$$

5

CONSTRUCTION OF THE RATIONAL COMPONENTS OF AN n-GON

A monogon has only one rational component, itself. So let $n \neq 1$.

In the Boolean algebra of cyclic n-gonal classes, let $\mathfrak{A}_n^*$ be the *class complementary to the Q-regular zero-point class* $\overset{\circ}{\mathfrak{R}}_n$. It is a free class. Since $\overset{\circ}{\mathfrak{R}}_n$ is the intersection of all classes $\mathfrak{A}_n^d$ with $d \| n$, and since $\mathfrak{A}_n^d$ is complementary to $\mathfrak{A}_{d,d}$ (chapter 4, theorem 1),

$$\mathfrak{A}_n^* = \sum_{d \| n} \mathfrak{A}_{d,d}. \tag{12}$$

2 See, e.g., B.H. Hasse, *Vorlesungen über Zahlentheorie* (Berlin, 1950), §4.7.

The class $\mathfrak{A}_n^*$ is thus the sum of all proper periodic classes; a reduction analogous to theorem 1 is possible. On the other hand, since the cyclic classes of theorem 3 form a direct decomposition of $\mathfrak{A}_n$, $\mathfrak{A}_n^*$ is the sum of all these classes except $\mathfrak{R}_n$.

Since Im $\varepsilon_n = \mathfrak{R}_n$, $\mathrm{Im}(1 - \varepsilon_n) = \mathfrak{A}_n^*$.

EXAMPLE $\mathfrak{A}_6^*$ is the class of prisms (see §5.5). $\mathfrak{A}_{12}^*$ is the class of 12-gons whose two omitting sub-6-gons are prisms.

Now let us consider the decomposition of an n-gon A into its rational components, written as

$$A = \sum_{d \| n} \varepsilon_d A + \varepsilon_n A = A^* + \varepsilon_n A \quad \text{with } A^* = \sum_{d \| n} \varepsilon_d A = (1 - \varepsilon_n)A. \tag{13}$$

Suppose that for every proper divisor d of n the proper periodic n-gon $\mu_d A$ is constructed, whose vertices are, by definition of μ_d, the centres of gravity of omitting subpolygons of A (chapter 4, theorem 1).

The components $\varepsilon_d A$ ($d \neq n$) can be built up from some of these proper periodic n-gons, namely from $\mu_d A$ and the n-gons $\mu_t A$ for $t \| d$, by theorem 7. But of these n-gons, only those for which d/t is square-free give a contribution. Similarly, on the basis of (11) the n-gon A^* can be constructed from some of the above n-gons $\mu_d A$. A^* and its summand $\varepsilon_d A$ ($d \| n$) lie in the class $\mathfrak{A}_n^*$.

The Q-regular component $\varepsilon_n A$ does not lie in the class $\mathfrak{A}_n^*$ (unless A already lies in $\mathfrak{A}_n^*$, in which case $\varepsilon_n A = O$) and is not built up out of proper periodic n-gons. But it can be constructed as $A - A^*$ as we have already seen in chapter 5, theorem 6, for $n = 6$. A and A^* are isobaric, and, on forming the difference, we obtain a Q-regular n-gon with centre of gravity o.

Exercises

1 If n is a power of a prime p, then the class complementary to the Q-regular zero-point class is a periodic class, namely the class of (n/p)-gons counted p times.

2 To each free cyclic class $\mathfrak{C}$ of p^l-gons we can associate two free classes of p^{l+1}-gons as follows: (1) the class of p^{l+1}-gons whose omitting p^l-gons lie in $\mathfrak{C}$; and (2) the class of p^{l+1}-gons whose omitting p^l-gons lie in $\mathfrak{C}$ and are isobaric. Can all free cyclic classes of p^{l+1}-gons be obtained in this way when $K = Q$?

11
Complex components of an *n*-gon

1

Let the field given in §1.1 contain an *n*th root of unity *w*. Then the set of all *n*-gons contains the set of *n*-gons

$$\{a, wa, w^2a, \ldots, w^{n-1}a\} \quad \text{with } a \in V. \tag{1}$$

These *n*-gons, each of whose vertices comes from the previous by multiplication with *w*, are called *w-n*-gons. If $a \neq o$, then all the vertices of (1) lie in the one-dimensional subspace Ka of V.

If $w = 1$, then the set of *w-n*-gons is the class of trivial *n*-gons. If $w \neq 1$, then *w* is a zero of the polynomial $(x^n - 1)/(x - 1) = 1 + x + x^2 + \ldots + x^{n-1}$, and thus

$$1 + w + w^2 + \ldots + w^{n-1} = 0 \quad \text{for } w \neq 1. \tag{2}$$

Thus if $w \neq 1$, every *w-n*-gon has centre of gravity o in V; for

$$(1/n)(a + wa + w^2a + \ldots + w^{n-1}a) = o \quad \text{for all } a \in V \text{ and } w \neq 1. \tag{3}$$

If *w* is in *K*, then the *n*th root of unity is w^{-1}. We obtain the w^{-1}-*n*-gons by taking the vertices of a *w-n*-gon in the reverse order. When $w \neq \pm 1$, only the zero *n*-gon O is a w^{-1}-*n*-gon as well as a *w-n*-gon.

The *w-n*-gons form a cyclic *n*-gonal class with degree of freedom 1, namely the atomic class defined by the prime divisor $x - w$ of $x^n - 1$. For, if $n \neq 1$, the *n*-gons (1) are the solutions of the cyclic system of equations $a_2 = wa_1, \ldots$. In the special case $n = 1$, the set of *w-n*-gons is the set of all monogons, and the assertion is equally valid (cf. §6.3).

If *K* contains all *n*th roots of unity, so that $x^n - 1$ factors in $K[x]$ into linear factors, then for every *n*th root of unity *w* there is a class of *w-n*-gons, and the following holds (§6.3 or chapter 8, theorem 5):

THEOREM 1 *If K contains the nth roots of unity, then the atomic cyclic n-gonal classes are the classes of w-n-gons (where w runs through all the nth roots of unity).*

If we add arbitrary trivial n-gons to the w-n-gons (1), we obtain the set $\mathfrak{C}_w$ of n-gons

$$(a+b,\ wa+b,\ w^2a+b,\ ...,\ w^{n-1}a+b) \quad \text{with } a, b \in V. \tag{4}$$

$\mathfrak{C}_1$ is again the class of trivial n-gons. If $w \neq 1$, the w-n-gons form a zero-point class by (3) and $\mathfrak{C}_w$ is the corresponding free cyclic class with degree of freedom 2; $\mathfrak{C}_w$ is the class defined by $(x-1)(x-w)$ and an upper neighbour of the class of trivial n-gons in the Boolean algebra of cyclic n-gonal classes.

Now let K contain the nth roots of unity. Then there are $\varphi(n)$ classes

$$\mathfrak{C}_w, \quad \text{with primitive } n\text{th root of unity } w. \tag{5}$$

The n-gons of these classes (all n-gons (4) where w is a primitive nth root of unity) are called *regular n-gons*. Since the classes (5) are free classes, all trivial n-gons are, in particular, regular.[1] A non-trivial regular n-gon consists of n distinct points.

The case $\varphi(n) = 1$, so that there is only one class (5), occurs only when $n = 1$ or 2. For $n = 1$ or 2, $\mathfrak{C}_1$ or $\mathfrak{C}_{-1}$ (respectively) is the only class (5) and already contains all n-gons; thus every monogon and every digon is regular (for any allowable K). The case $n = 1$ is particularly special in that the only class (5) is the class of trivial n-gons. For $n \neq 1$, all classes (5), since $w \neq 1$, are upper neighbours of the class of trivial n-gons in the Boolean algebra of cyclic classes.

The set of regular n-gons is by definition a union of cyclic classes, the classes (5). But adding n-gons from two different classes (5) can give us non-regular n-gons. Thus the set of regular n-gons is a cyclic class if and only if $\varphi(n) = 1$, that is, $n = 1$ or $n = 2$.

The sum of the classes (5) is the class of Q-regular n-gons. (The sum of the classes (5) is defined by the divisor $(x-1)F_n(x)$ of x^n-1 for $n \neq 1$ (chapter 8, theorem 5) and by $x-1$ for $n = 1$.) $n = 1, 2$ are also the only cases where every Q-regular n-gon is already regular.

COROLLARY TO THEOREM 1   *Let K contain the nth roots of unity. The class of trivial n-gons is an atomic cyclic class, and, for $n = 1$, the only class. If $n \neq 1$, there appear among the atomic cyclic classes $\varphi(n)$ classes of regular n-gons with centre of gravity o; their sum is the Q-regular zero-point class $\mathfrak{R}_n$. All further atomic cyclic classes are classes of regular d-gons with centre of gravity o, counted n/d times, for non-trivial divisors d of n.*

A proof of the last statement is contained in the following remarks.

For each point a there is, in each of the classes of theorem 1, exactly one n-gon with a as initial point, and it is suggestive to compare these n-gons with each

---

[1] Since the class of trivial n-gons exists for all n and K, it would be sensible to call the trivial n-gons regular even when K does not contain the nth roots of unity.

other. For this purpose let w be a primitive nth root of unity and $l \in \{1, 2, ..., n\}$. Then the classes of theorem 1 are the sets of w^l-n-gons. The w-n-gon with initial point a will be denoted by $A_1(a)$, and in general the w^l-n-gon with initial point a by $A_l(a)$.

$A_1(a)$ is a regular n-gon. The set of vertices of $A_l(a)$ is a subset of the set of vertices of $A_1(a)$. If w^l is a primitive nth root of unity, then the sets of vertices of $A_l(a)$ and $A_1(a)$ are equal, and $A_l(a)$ is a regular n-gon as well. $A_l(a)$ is obtained from $A_1(a)$ by running through the vertices of $A_1(a)$ in jumps[2] relatively prime to n. In general, if w^l is a primitive dth root of unity ($d \mid n$), then $A_l(a)$ is a regular d-gon, counted n/d times (the d-gon is the w^l-d-gon with initial point a). If $d \neq n$, $A_l(a)$ is properly periodic.

$A_n(a) = (a, a, ..., a)$ has centre of gravity a. On the other hand, if $n \neq 1$, $A_1(a), ..., A_{n-1}(a)$ are n-gons with centre of gravity o, by (3). If one of them is a d-gon counted several times, then the d-gon also has centre of gravity o.

From the above we obtain the following theorem which should clarify the concept of regularity:

THEOREM 2   *If A is a regular n-gon, then all n-gons which are obtained by running through the vertices of A in jumps relatively prime to n are regular. If d is a divisor of n, then the omitting sub-d-gons of a regular n-gon A are regular d-gons; for $d \neq 1$, they have the same centre of gravity as A.*

Exercises

1 The arithmetic mean of the n-gons $A_1(a), ..., A_n(a)$ is the n-gon $(a, o, ..., o)$.
2 Does the concept of Q-regularity have the properties mentioned in theorem 2?
3 A prime field of characteristic p with $p \nmid n$ contains the nth roots of unity if and only if $p \equiv 1 \bmod n$.

2
THE CASE OF THE COMPLEX FIELD

If K is the field C of complex numbers, then every nth root of unity w has a representation

$$w = e^{i\theta} = \cos \theta + i \sin \theta \quad \text{with } \theta = (2\pi/n)l \quad (l = 1, 2, ..., n).$$

Every one-dimensional subspace Ca of V ($a \neq o$) may be represented as a Gaussian number plane. In it

$$x \to wx$$

2 More clearly, for an n-gon $A = (a_1, a_2, ..., a_n)$ and l relatively prime to n, the n-gon $(a_1, a_{1+l}, a_{1+2l}, ..., a_{1+(n-1)l})$ has the same set of vertices as A and we say that it is obtained from A by *running through its vertices in jumps relatively prime to n* ($l = 1$ is allowed).

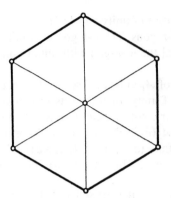

Figure 87

is a rotation about the origin through angle θ. Every vertex of a w-n-gon (1) comes from the preceding vertex by this rotation. *Thus our regular n-gons are regular n-gons of a Gaussian number plane, in the metric sense.* See figure 87.

In particular, for $n \neq 1$ the w-n-gons with $w = e^{i\theta}$, $\theta = 2\pi/n$ are regular (in the sense of our definition). These are 'positively' oriented regular n-gons with centre of gravity o in the usual sense (see, for example, figure 88); they form a cyclic class. The oppositely oriented n-gons, the associated w^{-1}-n-gons, also form a cyclic class. If $n > 2$, these two classes of regular n-gons with centre of gravity o are distinct and have only the zero n-gon in common. From our definition there are, when $\varphi(n) > 2$, still more regular n-gons with centre of gravity o: e.g., if $n = 5$, the class of positively oriented regular pentagrams with centre of gravity o (figure 89) and the class of oppositely oriented pentagrams.

The n-gons of the n atomic cyclic classes are:

for $n = 4$: the trivial 4-gons; the doubly counted digons $(a, -a, a, -a)$; the squares $(a, ia, -a, -ia)$; the oppositely oriented squares;

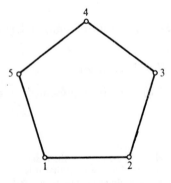

Figure 88

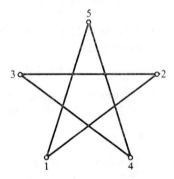

Figure 89

for $n = 5$ (let $\theta = \frac{2}{5}\pi$, $w = e^{i\theta}$): the trivial 5-gons, the regular 5-gons $(a, wa, ..., w^4 a)$; the oppositely oriented regular 5-gons; the regular pentagrams $(a, w^2 a, w^4 a, wa, w^3 a)$; the oppositely oriented regular pentagrams;

for $n = 6$ (let $\theta = \frac{2}{6}\pi$, $w = e^{i\theta}$): the trivial 6-gons; the digons counted three times $(a, -a, ..., -a)$; the doubly counted regular triangles $(a, w^2 a, w^4 a, a, w^2 a, w^4 a)$; the oppositely oriented, doubly counted regular triangles; the regular 6-gons $(a, wa, ..., w^5 a)$; the oppositely oriented regular 6-gons.

3

COMPLEX COMPONENTS OF AN n-GON

Let K contain the nth roots of unity. Every n-gon A can be represented uniquely as a sum of n-gons from the atomic cyclic classes and hence by theorem 1 as a sum of w-n-gons where w runs through the nth roots of unity. We call the term of A belonging to a given w the w-*components of* A and all these components of A the *complex components of* A.

To find the complex components for an n-gon $A = (a_1, a_2, ..., a_n)$ we proceed as follows:

First, write

$$(a_1, a_2, ..., a_n) = (a_1, 0, ..., 0) + (0, a_2, ..., 0) + ... + (0, 0, ..., a_n). \tag{6}$$

Since the sum of all the nth roots of unity is zero for $n \neq 1$ (see (2)), we have the identity

$$(a_1, 0, ..., 0) = \sum_{\text{all } w} (1/n)(a_1, wa_1, w^2 a_1, ..., w^{n-1} a_1). \tag{7}$$

Thus $(a_1, 0, ..., 0)$ is represented as a sum of n-gons from the atomic classes of theorem 1 – the w-n-gons with initial point $(1/n)a_1$. Because of the uniqueness of the atomic decomposition these terms are the complex components of $(a_1, 0, ..., 0)$. Correspondingly we have

$$(0, a_2, 0, ..., 0) = \sum_{\text{all } w} (1/n)(w^{n-1} a_2, a_2, wa_2, ..., w^{n-2} a_2),$$

and the complex components of $(0, a_2, 0, ..., 0)$ appear on the right-hand side. Continue thus. For any w the w-component of A is obtained by adding the w-components of the terms in (6). A w-n-gon results, the w-n-gon with initial point

$$(1/n)(w^n a_1 + w^{n-1} a_2 + ... + wa_n) = (1/n)\sum_{i=0}^{n-1} w^{n-i} a_{i+1}. \tag{8}$$

This gives the desired solution.

(This result can also be interpreted as follows: for each vertex a_i of A form the w^{-1}-n-gon with initial point a_i and write these n n-gons one under the other. Then the centre of gravity of the points on the main diagonal of this scheme is

F

the first vertex (8) of the *w*-component of A. Further vertices of this component are obtained by multiplying the first one by w, w^2, ..., w^{n-1}; but they are also the centres of gravity of the 'parallels' to the main diagonal.)

The *w-n*-gon with initial point (8) is the image of $(a_1, a_2, ..., a_n)$ under the cyclic mapping with coefficient *n*-tuple $(1/n)(1, w^{-1}, w^{-2}, ..., w^{-(n-1)})$ i.e., by chapter 2, theorem 3, the cyclic mapping

$$e_w(\zeta): = (1/n)(1 + w^{-1}\zeta + w^{-2}\zeta^2 + ... + w^{-(n-1)}\zeta^{n-1}) = (1/n)\sum_{i=0}^{n-1} w^{-i}\zeta^i.$$

The result can therefore be expressed as follows:

THEOREM 3   *If K contains the nth roots of unity, then for every n-gon A*

$$A = \sum_{\text{all } w} e_w(\zeta)A \quad \text{with} \quad e_w(\zeta) = (1/n)\sum_{i=0}^{n-1} w^{-i}\zeta^i$$

is the unique decomposition of A into n-gons of the atomic cyclic classes.

The term $e_w(\zeta)A$ is a *w-n*-gon, and from theorem 3 we have

COROLLARY $e_w(\zeta)$ *projects the set of all n-gons onto the (atomic) cyclic class defined by* $x - w$, *the class of w-n-gons.*

This corollary shows that, if K contains the *n*th roots of unity, then the $e_w(\zeta)$ are the atomic cyclic projections. Thus, conversely, theorem 3 can be proved from the corollary. But the corollary can also be verified by a simple calculation (rather like the first two statements from chapter 4, theorem 1). The only interesting point is how we get the mapping $e_w(\zeta)$, and, let us say in addition, how $e_w(\zeta)$ and the corollary come from the general theorems of chapter 8. Chapter 8 gives two procedures for finding cyclic projections which map the set of all *n*-gons onto the cyclic class defined by $x - w$:

1 By chapter 8, (15), $e_t(\zeta)$ is the desired projection for $t(x) = x - w$. If we calculate first the polynomial $e_t(x)$ for this $t(x)$ using the derivative formula from chapter 8, theorem 3, then we obtain the polynomial

$$e_w(x) = (1/n)(1 + w^{-1}x + (w^{-1}x)^2 + ... + (w^{-1}x)^{n-1}) = (1/n)\sum_{i=0}^{n-1} w^{-i}x^i$$

and from this we get our $e_w(\zeta)$ by substituting ζ.

2 By chapter 8, theorem 8 (a), we obtain the desired projection by substituting ζ in each polynomial of $K[x]$ whose value is 1 at w and 0 at the remaining *n*th roots of unity. A polynomial of degree $n - 1$ which is not equal to 0 at w and equal to 0 at all *n*th roots of unity $\neq w$ is the divisor of $x^n - 1$ complementary to $x - w$. Thus we obtain a polynomial which is equal to 1 at w and 0 at the remaining *n*th roots of unity by dividing the polynomial

$$(x^n - 1)/(x - w) = w^{-1}(1 + (w^{-1}x) + (w^{-1}x)^2 + ... + (w^{-1}x)^{n-1})$$

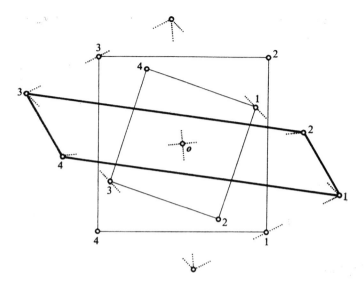

Figure 90

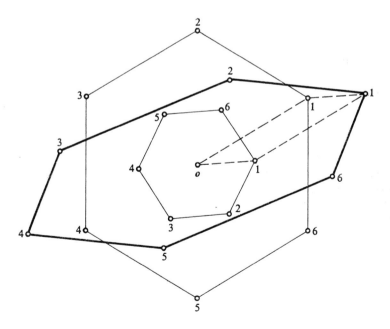

Figure 91

by its functional value at w, that is, by $w^{-1}n$. The resulting polynomial is $e_w(x)$. In this way we again obtain $e_w(\zeta)$ and the corollary.

It is especially interesting to decompose a Q-regular n-gon with centre of gravity o into its complex components; all these components are regular n-gons with centre of gravity o.

EXAMPLE   In a Gaussian number plane (the case $V = K = C$), the following holds:
Every parallelogram with centre of gravity o can be represented uniquely as the sum of two oppositely oriented squares with centre of gravity o.
    In figure 90 $A = (a_1, a_2, -a_1, -a_2)$ decomposes into the i-component of A

$$\tfrac{1}{2}(a_1 - ia_2, ia_1 + a_2, -a_1 + ia_2, -ia_1 - a_2)$$

and the $(-i)$ component of A:

$$\tfrac{1}{2}(a_1 + ia_2, -ia_1 + a_2, -a_1 - ia_2, ia_1 - a_2).$$

Correspondingly every triangle with centre of gravity o is uniquely representable as the sum of two oppositely oriented regular triangles with centre of gravity o, and every affinely regular hexagon with centre of gravity o as the sum of two oppositely oriented regular hexagons with centre of gravity o (figure 91).

Exercises

1 How is the decomposition of a parallelogram into two squares with the same centre of gravity specialized when the parallelogram is a rhombus?
2 For which n can all complex components of an n-gon be constructed with ruler and compass in a Gaussian number plane?
3 Let K contain the nth roots of unity. The nth roots of unity are the eigenvalues of the endomorphism ζ of $\mathfrak{A}_n$; the class of w-n-gons is the eigenspace belonging to the eigenvalue w (the set of A's with $\zeta A = wA$); the atomic decomposition of $\mathfrak{A}_n$ is the decomposition into the eigenspaces of ζ; $e_w(\zeta)$ is the projection of $\mathfrak{A}_n$ onto the eigenspace belonging to w.
4 Every cyclic mapping $f(\zeta)$ acts on the class of w-n-gons like the stretching with ratio $f(w)$.

12
The real components of an *n*-gon

1

Assuming again the data from §1.1, let us consider the following mapping in the vector space of *n*-gons:

$$A = (a_1, a_2, ..., a_n) \to A^* = (a_1, a_n, ..., a_2) \tag{1}$$

which maps any *n*-gon *A* onto the *n*-gon *A** with the same initial point but with vertices taken in the opposite order. Then $A^{**} = A$, $(A+B)^* = A^*+B^*$, $(cA)^* = cA^*$: (1) is an involutory linear mapping of $\mathfrak{A}_n$ onto itself. Moreover $(\zeta A)^* = \zeta^{-1}A^*$, and thus for every cyclic mapping $f(\zeta)$

$$(f(\zeta)A)^* = f(\zeta^{-1})A^*. \tag{2}$$

Under (1) every cyclic class is mapped onto a cyclic class: to be precise, the class Ker $f(\zeta)$ onto the class Ker $f(\zeta^{-1})$. A cyclic class is called *anticyclic* if (1) maps it onto itself; this is the case if, when a cyclic class contains an *n*-gon *A*, it always contains the *n*-gon *A**.

THEOREM 1 *The anticyclic cyclic n-gonal classes form a Boolean subalgebra of the Boolean algebra of all cyclic n-gonal classes. The subalgebra contains the four basic classes. If it contains a free cyclic class, it contains the associated zero-point class and conversely.*

The last statement comes from chapter 3, theorem 2.
In $K[x]$, we consider the mapping

$$f(x) \to f^*(x) = x^{\deg f} f(x^{-1}) \tag{3}$$

for polynomials $\neq 0$. Then $c^* = c$ for each $c \neq 0$ in *K*, and $(f(x)g(x))^* = f^*(x)g^*(x)$. $f(x)$ is called *symmetric* if $f^*(x) = f(x)$, and *antisymmetric* if $f^*(x) = -f(x)$.

(i) *If $f^*(x)$ is an associate of $f(x)$, then $f(x)$ is symmetric or antisymmetric (and conversely).*

PROOF Let $f^*(x) = cf(x)$ with $c \neq 0$ in K. If $c_m \neq 0$ is the highest coefficient and c_0 the constant term of $f(x)$, then $c_0 = cc_m$, $c_m = cc_0$. Thus $c_m = c^2 c_m$, and hence $c = \pm 1$.

(ii) *Under (1) the cyclic class* $\operatorname{Ker} f(\zeta)$ *is mapped onto the cyclic class* $\operatorname{Ker} f^*(\zeta)$.

PROOF Since the powers of ζ are units in $K[\zeta]$, $f(\zeta^{-1})$ and $\zeta^{\deg f} f(\zeta^{-1})$ are associates in $K[\zeta]$ and so have the same kernel (see chapter 6, theorem 5): $\operatorname{Ker} f(\zeta^{-1}) = \operatorname{Ker} f^*(\zeta)$.

In the following discussion we shall make frequent use of our standard isomorphism of the lattice of divisors of $x^n - 1$ onto the Boolean algebra of cyclic n-gonal classes (chapter 8, theorem 5). This isomorphism is defined by

$$t(x) \to \operatorname{Ker} t(\zeta) \tag{4}$$

and we prove in addition

THEOREM 2 *Let* $t(x) \mid x^n - 1$. *If* $t(x)$ *is symmetric or antisymmetric, then* $\operatorname{Ker} t(\zeta)$ *is anticyclic and conversely.*

First we note that $t(x) \mid x^n - 1$ implies $t^*(x) \mid x^n - 1$. For, if $x^n - 1 = t(x)\bar{t}(x)$, then $-(x^n - 1) = (x^n - 1)^* = t^*(x)\bar{t}^*(x)$.

PROOF OF THEOREM 2 The isomorphism (4) shows that, if $t(x), t^*(x)$ are associates, then $\operatorname{Ker} t(\zeta) = \operatorname{Ker} t^*(\zeta)$ and conversely. (For the converse part use the previous remark.) The assertion then follows from (i) and (ii).

Obviously $x - 1$ is antisymmetric, and, if $f(x)$ is symmetric, $(x - 1)f(x)$ is also antisymmetric. If char $K \neq 2$, then we obtain all antisymmetric polynomials from the symmetric ones by multiplying them with $x - 1$.

(iii) *For char* $K \neq 2$, *the following holds: the set of antisymmetric polynomials is the set of all polynomials* $(x - 1)f(x)$ *where* $f(x)$ *is symmetric.*

PROOF For each polynomial $g(x)$, $g^*(1) = g(1)$. If $g(x)$ is antisymmetric, then $g(1) = -g(1)$, and thus, for char $K \neq 2$, $g(1) = 0$. Hence $g(x)$ is divisible by $x - 1$ and so representable in the form $g(x) = (x - 1)f(x)$. Since $g^*(x) = -g(x)$, $(x - 1)^* f^*(x) = -(x - 1)f(x)$, and thus $f^*(x) = f(x)$.

Using only the assumptions of §1.1 we can prove:

(iv) *If a divisor of the symmetric polynomial*

$$P_n(x) = (x^n - 1)/(x - 1) = 1 + x + x^2 + \ldots + x^{n-1}$$

is either symmetric or antisymmetric, it must be symmetric.

PROOF $P_n(1) = n \cdot 1 \neq 0$. Therefore $P_n(x)$ and so its divisors are not divisible by $x-1$. If char $K \neq 2$, then $P_n(x)$ possesses no antisymmetric divisors by (iii). If char $K = 2$, then every antisymmetric polynomial is symmetric and the assertion is trivial.

COROLLARY TO THEOREM 2   *The symmetric divisors of $P_n(x)$ define the anticyclic zero-point classes; the symmetric divisors of $P_n(x)$, multiplied by $x-1$, define the free anticyclic classes.*

PROOF   The anticyclic zero-point classes are the anticyclic cyclic classes which are included in the zero isobaric class. Since the latter is defined by $P_n(x)$ (see §8.4), the anticyclic zero-point classes are defined by the divisors of $P_n(x)$ which are symmetric or antisymmetric (theorem 2). Then by (iv) these are the symmetric divisors of $P_n(x)$.

If we multiply the defining polynomials of the anticyclic zero-point classes by $x-1$, then we obtain the defining polynomials of the associated free cyclic classes (see §8.4). By theorem 1 these classes are the free anticyclic classes.

EXAMPLE   The cyclotomic polynomials $F_d(x)$ with $d \mid n$ are symmetric if $d \neq 1$ (see appendix 2), while $F_1(x)$ is antisymmetric. Now let K be the rational number field Q. Then these polynomials define the atomic cyclic classes; thus by theorem 2 the atomic cyclic classes are anticyclic, and by theorem 1 all cyclic classes are anticyclic.

Exercises

1 Let $n \neq 1, 2$ and w be a primitive nth root of unity in K. (1) interchanges the class of w-n-gons and the class of w^{-1}-n-gons. The regular n-gonal classes are not anticyclic.
2 Over the Gaussian number field $Q(i)$ only 8 of the 16 cyclic quadrangular classes are anticyclic, namely those which already exist over Q.
3 For char $K \neq 2$ the following holds: the symmetric divisors of $x^n - 1$ define the anticyclic zero-point classes; the antisymmetric divisors of $x^n - 1$ define the free anticyclic classes.

2
A SPECIAL TYPE OF CYCLIC SYSTEMS OF EQUATIONS

With a view to finding the finest possible decomposition of the vector space of n-gons into anticyclic cyclic classes (which in general will be possible only over special fields), we shall consider in this section and the next anticyclic cyclic n-gonal classes of small degree, namely *the class of trivial n-gons and the anticyclic zero-point classes of degrees 1 and 2.* (5)

The class of trivial n-gons is the only free cyclic class of degree 1; it is anticyclic. There exists an anticyclic zero-point class of degree 1 only when n is even, say $n = 2m$: the class $\mathfrak{A}_{2,m}$ of 2-gons with centre of gravity o, counted m times, defined by $x+1$. Of the classes (5), only the anticyclic zero-point classes of degree 2 require a closer study, and these can exist only for $n \geq 3$. They are defined by the quadratic symmetric divisors of the polynomial $P_n(x)$ (corollary to theorem 2 and chapter 9, theorem 5), and thus, without loss of generality, by the polynomials $x^2 - cx + 1$ in $K[x]$ with

$$x^2 - cx + 1 \mid P_n(x). \tag{6}$$

The classes (5), if they exist, are atomic classes of the Boolean algebra of anticyclic cyclic n-gonal classes.[1]

Let $n \geq 3$ and $x^2 - cx + 1$ be a quadratic symmetric polynomial of $K[x]$; at first we do not assume that it divides $P_n(x)$. Its coefficient n-tuple $(1, -c, 1, 0, ..., 0)$ defines the cyclic system of equations

$$a_1 - ca_2 + a_3 = o, \ldots \tag{7}$$

and with it an *anticyclic cyclic class* $\mathfrak{C}(c)$ *whose degree* ≤ 2. The n-gons of this class are at most two-dimensional. Moreover $\mathfrak{C}(c) = \text{Ker}(\zeta^2 - c\zeta + 1)$.

The geometric meaning of an n-gon which satisfies a cyclic system of equations (7) is as follows: *there exists some* $c \in K$ *such that the sum of* a_1, a_3 *is c times* a_2, *the sum of* a_2, a_4 *is c times* a_3, *etc.*

If $(a_1, a_2, ..., a_n)$ is an n-gon of the class $\mathfrak{C}(c)$, then its vertices may be written as linear combinations of a_1, a_2: if we define a sequence $\hat{c}_0, \hat{c}_1, \hat{c}_2, \ldots$ of elements of K for a given c by

$$\hat{c}_0 = -1, \quad \hat{c}_1 = 0, \quad \hat{c}_{k+2} = -\hat{c}_k + c\hat{c}_{k+1} \quad (k \geq 0), \tag{8}$$

then from the first $n-2$ equations of the system (7) we have

$$a_k = -\hat{c}_{k-1}a_1 + \hat{c}_k a_2 \quad \text{for } k = 1, 2, ..., n. \tag{9}$$

The two last equations of the system (7) can be written in the form

$$a_1 = -\hat{c}_n a_1 + \hat{c}_{n+1} a_2, \qquad a_2 = (\hat{c}_{n-1} + c)a_1 - \hat{c}_n a_2 \tag{10}$$

using the expressions for a_{n-1} and a_n from (9).

If $\hat{c}_n = -1$, $\hat{c}_{n+1} = 0$, then $\hat{c}_{n+k} = \hat{c}_k$ for all $k \geq 0$. We then say that the sequence (8) is *periodic with period n*. (If this is the case, then note that $\hat{c}_{n-1} = -c$.)

---

1 An anticyclic zero-point class of degree 1 exists only if $n = 2m$ as mentioned earlier; it is defined by $x+1$ and is not contained in an anticyclic zero-point class of degree 2, since $x+1 \mid x^2 - cx + 1 \mid P_n(x)$ is impossible. $(x+1 \mid x^2 - cx + 1$ implies $c = -2$ but $x^2 + 2x + 1 = (x+1)^2$ is not a divisor of the square-free polynomial $P_n(x)$.)

EXAMPLE   The sequence (8) is, for

$$c = 2: \quad -1, 0, 1, 2, 3, 4, \ldots ;$$
$$c = -2: \quad -1, 0, 1, -2, 3, -4, \ldots ;$$
$$c = -1: \quad -1, 0, 1, -1, 0, \ldots ;$$
$$c = 0: \quad -1, 0, 1, 0, -1, 0, \ldots ;$$
$$c = 1: \quad -1, 0, 1, 1, 0, -1, -1, 0, \ldots .$$

$\mathfrak{C}(2)$ is the class of trivial n-gons;

$\mathfrak{C}(-2)$ is the class $\mathfrak{A}_{2,m}$ if $n = 2m$;

$\mathfrak{C}(-1)$ is the class of triangles with centre of gravity o if $n = 3$;

$\mathfrak{C}(0)$ is the class of parallelograms with centre of gravity o if $n = 4$;

$\mathfrak{C}(1)$ is the class of affinely regular hexagons with centre of gravity o if $n = 6$.

THEOREM 3   *Let $n \geq 3$. The cyclic n-gonal classes, distinct from the zero class, which are described by cyclic systems of equations (7), are the class of trivial n-gons and the anticyclic zero-point classes of degrees 1 and 2. The system of equations (7) defines an anticyclic zero-point class of degree 2 if and only if the sequence (8) is periodic with period n.*

PROOF   The anticyclic zero-point classes of degree 2 are precisely those classes $\mathfrak{C}(c)$ for which c satisfies condition (6). A free anticyclic cyclic class of degree 2 exists only for even $n = 2m$, that is, the class $\mathfrak{A}_{2,m}$ defined by $(x-1)(x+1)$ (see theorem 1); it is not represented by a cyclic system of equations (7). If (6) does not hold, $\mathfrak{C}(c)$ is therefore an anticyclic cyclic class of degree < 2, i.e., the zero class or the class $\mathfrak{A}_{1,n} = \mathfrak{C}(2)$ of trivial n-gons or, if $n = 2m$, the class $\mathfrak{A}_{2,m} = \mathfrak{C}(-2)$.

As for the second statement of theorem 3, $\deg \mathfrak{C}(c) = 2$ means that the system (7) can be satisfied by arbitrarily chosen a_1, a_2. Since, from the first $n-2$ equations of the system, $a_3, \ldots, a_n$ are uniquely determined by a_1, a_2 as seen in (9), this is precisely the case when the two equations (10) hold for arbitrary a_1, a_2. For this it is necessary and sufficient that, in the two equations (10), each of a_1, a_2 has equal coefficients on both sides, that is

$$1 = -\hat{c}_n, \quad 0 = \hat{c}_{n+1}, \quad 0 = \hat{c}_{n-1} + c, \quad 1 = -\hat{c}_n. \tag{11}$$

(If equations (10) hold for arbitrary a_1, a_2, then in particular they hold for pairs $a_1 \neq 0, a_2 = 0$ and $a_1 = 0, a_2 \neq 0$, which give us (11). On the other hand, if (11) holds, equations (10) are valid for any a_1, a_2.) Equations (11) say that the sequence (8) is periodic with period n.

The second statement of theorem 3 gives a meaning to condition (6): *condition (6) is equivalent to the fact that the sequence (8) associated with the element c is periodic with period n*. Thus such periodic sequences of elements from K are connected with the anticyclic zero-point classes of degree 2.

Exercises

1 Let (6) hold. In a splitting field of $x^n - 1$ over K, $x^2 - cx + 1$ has as zeros two nth roots of unity which are inverses to each other: w, w^{-1}. Then $c = w + w^{-1}$ and, because of (6), $w = \pm 1$. From this prove that the sequence (8) is periodic with period n.

2 If sequence (8) is periodic with period n, then
$$\hat{c}_0 + \hat{c}_1 + \hat{c}_2 + \dots + \hat{c}_{n-1} = 0.$$

3 For which rational numbers c, and for which real numbers c, is the sequence (8) periodic?

4 If $c \in K$, if a_1, a_2 are linearly independent elements of V (here we must have dim $V \geq 2$), and if we define recursively $a_k = -a_{k-2} + ca_{k-1}$ for $k = 3, 4, \dots$, then the sequence of points $a_1, a_2, a_3, \dots$ is periodic with period n if and only if sequence (8) is periodic with period n.

3
AFFINELY REGULAR n-GONS

Anticyclic zero-point classes of degree ≤ 2 which are contained in the Q-regular zero-point class $\mathfrak{R}_n$ are called *affinely regular zero-point classes*. (The only affinely regular zero-point class of degree 0 is the zero class. There exists an affinely regular zero-point class of degree 1 only for $n = 2$: the class $\mathfrak{A}_2 = \mathfrak{R}_2$ of digons with centre of gravity o.) Free cyclic classes associated with affinely regular zero-point classes are called *free affinely regular classes*. An n-gon is called *affinely regular* if it lies in an affinely regular class.

All affinely regular n-gons are Q-regular. They are at most two-dimensional and thus always plane figures. All trivial n-gons and all 1-gons, 2-gons, and 3-gons are affinely regular.

Since the Q-regular zero-point class is defined by the nth cyclotomic polynomial $F_n(x)$ for $n \neq 1$ (chapter 10, theorem 2), we have:

THEOREM 4   *For $n \geq 3$ the affinely regular zero-point classes $\neq \{O\}$ are the cyclic classes which are defined by quadratic symmetric divisors of $F_n(x)$.*

The affinely regular 4-gons are parallelograms; for $n = 6$ the definition coincides with the concept of affinely regular 6-gons introduced in chapter 1. For the cases $n = 1, 2, 3, 4, 6$ all Q-regular n-gons are affinely regular.

THEOREM 5   *There are non-trivial affinely regular n-gons over all fields K (with char $K \nmid n$) only for $n = 2, 3, 4, 6$.*

For even over the field of rational numbers the cyclotomic polynomial $F_n(x)$ with $n \geq 3$ has a square divisor only if it is a quadratic itself, i.e., if $\varphi(n) = 2$.

If K contains the nth roots of unity, then a quadratic symmetric divisor of $F_n(x)$, normalized to be monic, has the form

$$x^2 - (w + w^{-1})x + 1 = (x - w)(x - w^{-1}) \tag{12}$$

(w a primitive nth root of unity).

Then the isomorphism (4) shows:

THEOREM 6   *Let $n \geq 3$, and K contain the nth roots of unity. Then every affinely regular n-gon with centre of gravity o is uniquely representable as a sum of a regular w-n-gon and a regular w^{-1}-n-gon (w a primitive nth root of unity).*

The examples of §11.3 are illustrations of this.

LEMMA   *Let $n \geq 3$ and $x^2 - cx + 1$ be a quadratic symmetric divisor of $F_n(x)$ in $K[x]$:*

$$x^2 - cx + 1 \mid F_n(x), \tag{13}$$

and let $c_0, c_1, c_2, \ldots$ be a sequence of elements of K defined by

$$c_0 = 2, \quad c_1 = c, \quad c_{k+2} = -c_k + cc_{k+1} \quad (k \geq 0). \tag{14}$$

Then in $K[x]$

$$F_n(x) = \prod_{\substack{(k,n)=1 \\ k < n/2}} (x^2 - c_k x + 1). \tag{15}$$

PROOF   In a splitting field of $x^n - 1$ over K, c has a representation $c = w + w^{-1}$, where w is a primitive nth root of unity. Set

$$c_k' = w^k + w^{-k} \quad \text{for } k = 0, 1, 2, \ldots.$$

Then $c_0' = 2$, $c_1' = c$, and we have the recursion formula

$$c_{k+2}' = -c_k' + cc_{k+1}',$$

and thus $c_k' = c_k$. Since $F_n(x)$ is the product of the $(x - w^k)$'s with $(k, n) = 1$, we obtain the assertion by combining the factors $x - w^k$, $x - w^{-k}$.

As the proof of the lemma shows, *the sequence* (14) *is periodic with period n and has no smaller period.* Moreover $c_k = c_{n-k}$. If $n = 2m$ is even, $c_m = -2$.

From the lemma we have:

THEOREM 7   *Let $n \geq 3$. If $F_n(x)$ in $K[x]$ has a quadratic symmetric divisor, then $F_n(x)$ factors in $K[x]$ into quadratic symmetric factors. Non-trivial affinely regular n-gons exist over K (if and) only if $F_n(x)$ factors over K into quadratic symmetric factors.*

If, for a given $n \geq 3$ and K, there exists an affinely regular zero-point class $\neq \{O\}$, then there are $\frac{1}{2}\varphi(n)$ such classes; their sum is the Q-regular zero-point class. There are also non-trivial affinely regular d-gons for each divisor $d \neq 1$ of n. This can be proved, using theorem 7, from the fact that under the assumptions of the lemma

$$F_d(x) = \prod_{\substack{(k,n)=n/d \\ k < n/2}} (x^2 - c_k x + 1) \quad \text{for } d \,|\, n, \, d \geq 3. \tag{16}$$

(For $d = 2$ there is nothing to prove.)

These statements already hint at how the lemma gives a procedure by which we can associate a certain sequence of cyclic classes to an affinely regular zero-point class; this phenomenon should be discussed somewhat more systematically.

Let $n \geq 3$ and $\mathfrak{C}(c)$ be an affinely regular zero-point class which is defined by a quadratic symmetric divisor $x^2 - cx + 1$ of $F_n(x)$. For the sequence of elements c_k ($k = 1, 2, \ldots, n$) of the lemma form the sequence of cyclic n-gonal classes

$$\mathfrak{C}(c) = \mathfrak{C}(c_1), \, \mathfrak{C}(c_2), \, \ldots, \, \mathfrak{C}(c_n). \tag{17}$$

Then $\mathfrak{C}(c_n) = \mathfrak{C}(2) = \mathfrak{A}_{1,n}$ is the class of trivial n-gons. For $n = 2m$, $\mathfrak{C}(c_m) = \mathfrak{C}(-2) = \mathfrak{A}_{2,m}$. Except for these two special cases $k = n$ and $n = 2m$, $k = m$, $x^2 - c_k x + 1$ is a divisor of $P_n(x)$ (see the proof of the lemma), and therefore $\mathfrak{C}(c_k)$ is an anticyclic zero-point class of degree 2. $\mathfrak{C}(c_k) = \mathfrak{C}(c_{n-k})$. For odd n the different classes (17) are represented by $\mathfrak{C}(c_1), \mathfrak{C}(c_2), \ldots, \mathfrak{C}(c_{(n-1)/2}), \mathfrak{C}(c_n)$, and for even $n = 2m$ by $\mathfrak{C}(c_1), \mathfrak{C}(c_2), \ldots, \mathfrak{C}(c_m), \mathfrak{C}(c_n)$.

By theorem 4 and the lemma $\mathfrak{C}(c_k)$ is an affinely regular zero-point class if and only if $(k, n) = 1$.

In general, if $n/(k, n)$ is set equal to d, then $\mathfrak{C}(c_k)$ consists of affinely regular d-gons counted n/d times which for $k \neq n$ have centre of gravity o. We already know this for $k = n$ and, if $n = 2m$, for $k = m$. In all other cases the assertion follows from the fact that $x^2 - c_k x + 1$ is a divisor of $F_d(x)$, by (16) (cf. the lemma from §8.4).

Our goal now is to establish a geometric relation between the class $\mathfrak{C}(c)$ and the classes (17). For this let us consider the mappings

$$A = (a_1, a_2, \ldots, a_n) \rightarrow A_k = (a_1, a_{1+k}, a_{1+2k}, \ldots, a_{1+n-k}) \tag{18}$$

of $\mathfrak{A}_n$ into itself for $k = 1, 2, \ldots, n$. Note first that $A = A_1$, all A_k have initial point a_1, A_n is the trivial n-gon $(a_1, a_1, \ldots, a_1)$, and A_k and A_{n-k} differ only in their orientation: $A_{n-k} = A_k{}^*$. The A_k with $(k, n) = 1$ are the n-gons which are obtained by running through the vertices of A in jumps relatively prime to n. When $(k, n) = n/d$, and in particular when $k = n/d$, A_k is a sub-d-gon of A counted n/d times.

The mapping $A \rightarrow A_k$ associates with each n-gon of the affinely regular zero-point class $\mathfrak{C}(c)$ an n-gon of the class $\mathfrak{C}(c_k)$. In order to show this, we first prove:

COROLLARY TO THE LEMMA   *Under the assumptions of the lemma,*

$$x^2 - cx + 1 \mid x^{2k} - c_k x^k + 1 \quad \text{for } k = 1, 2, \ldots, n.$$

PROOF   In a splitting field of $x^n - 1$ over K, let $x^2 - cx + 1$ again be of the form (12). Then $c_k = w^k + w^{-k}$, and thus $w^{2k} - c_k w^k + 1 = 0$ and $w^{-2k} - c_k w^{-k} + 1 = 0$. The roots w, w^{-1} of $x^2 - cx + 1$ are, thus, also roots of $x^{2k} - c_k x^k + 1$. Therefore the assertion holds over an extension field of K, and also in $K[x]$.

This corollary shows that the given class $\mathfrak{C}(c) = \text{Ker}(\zeta^2 - c\zeta + 1)$ is contained in the class $\text{Ker}(\zeta^{2k} - c_k \zeta^k + 1)$. If $A = (a_1, a_2, \ldots, a_n) \in \mathfrak{C}(c)$, then $a_1 - c_k a_{1+k} + a_{1+2k} = o$, $\ldots$, that is $A_k \in \mathfrak{C}(c_k)$.

Thus the mappings $A \to A_k$ map the affinely regular n-gon A with centre of gravity o onto affinely regular n-gons with centre of gravity o, affinely regular d-gons with centre of gravity o counted several times (with non-trivial divisors d of n), and, for $k = n$, a trivial n-gon. If A is in the class $\mathfrak{C}(c)$, then the image n-gons are distributed among the classes $\mathfrak{C}(c_k)$. If A_k is in the class $\mathfrak{C}(c_k)$, then so is $A_{n-k} = A_k^*$, since the class is anticyclic.

We then realize that the following holds, analogous to theorem 2 of chapter 11:

THEOREM 8   *If A is an affinely regular n-gon, then so are all n-gons which are obtained by running through the vertices of A in jumps relatively prime to n. If d is a divisor of n, then the omitting sub-d-gons of A are affinely regular d-gons which, if $d \neq 1$, have the same[2] centre of gravity as A.*

For given n and K let there exist in $K[x]$ a quadratic symmetric divisor $x^2 - cx + 1$ of $F_n(x)$ (for $n \geq 3$). Then, using the notation of the lemma, the following holds:

$$x^n - 1 = \begin{cases} (x-1) \displaystyle\prod_{k=1}^{(n-1)/2} (x^2 - c_k x + 1) & \text{for odd } n, \\[3mm] (x-1)(x+1) \displaystyle\prod_{k=1}^{m-1} (x^2 - c_k x + 1) & \text{for even } n = 2m. \end{cases} \tag{19}$$

(Cf. (15), (16), and chapter 6, (8). For $n = 1, 2$, (19) is trivially true; but the factors $x^2 - c_k x + 1$, defined only for $n \geq 3$, are lacking.)

With the aid of the isomorphism (4), (19) gives the following decomposition of the vector space $\mathfrak{A}_n$ of n-gons:

$$\mathfrak{A}_n = \begin{cases} \mathfrak{A}_{1,n} \oplus \mathfrak{C}(c_1) \oplus \mathfrak{C}(c_2) \oplus \ldots \oplus \mathfrak{C}(c_{(n-1)/2}) & \text{for odd } n, \\[2mm] \mathfrak{A}_{1,n} \oplus \mathfrak{A}_{2,m} \oplus \mathfrak{C}(c_1) \oplus \mathfrak{C}(c_2) \oplus \ldots \oplus \mathfrak{C}(c_{m-1}) & \text{for even } n = 2m. \end{cases} \tag{20}$$

2 The equality of the centres of gravity follows from the Q-regularity of A.

Since a factorization of $(x^n-1)/(x-1)$ or of $(x^n-1)/(x-1)(x+1)$ into quadratic symmetric factors is uniquely determined up to associates, the classes $\mathfrak{C}(c_k)$ ($k = 1, 2, ..., (n-1)/2$ or $(n/2)-1$ respectively), named in (20), are all the anticyclic zero-point classes of degree 2. *Altogether the classes appearing in the decomposition* (20) *are the class of trivial n-gons and the anticyclic zero-point classes of degrees* 1 *and* 2; *and, if n ≥ 3, also the cyclic n-gonal classes ≠ {O} which are described by cyclic systems of equations of type* (7) (*theorem* 3).

Exercise

Let $n \leq 3$ and $\mathfrak{C}(c)$ be an affinely regular zero-point class. Then the mapping $A \to A_n$ (or $A \to A_{n-k}$) is a mapping of $\mathfrak{C}(c)$ onto $\mathfrak{C}(c_k)$ ($k = 1, 2, ..., n$).

4

THREE EXTREME CASES FOR THE BOOLEAN ALGEBRA OF CYCLIC n-GONAL CLASSES

Let n be a natural number. For any field K and any K-vector space V which satisfies the conditions in §1.1, the cyclic n-gonal classes may be defined; by the main theorem of §6.2 they form a Boolean algebra which we shall now denote by $L_n(K, V)$. The fundamental isomorphism (4) of the lattice of divisors of x^n-1 defined in $K[x]$ onto $L_n(K, V)$ exists for each V; it follows that the Boolean algebras $L_n(K, V)$ for a fixed K are isomorphic. In particular, the number of atomic classes of $L_n(K, V)$ is equal to the number of prime factors of x^n-1 in $K[x]$, and thus depends only on K; let it be denoted by $k_n(K)$. The number of cyclic n-gonal classes, the cardinality of $L_n(K, V)$, is $2^{k_n(K)}$, independent of the choice of V.

If $n_1, n_2, ..., n_{k_n(K)}$ are the degrees of the prime factors of x^n-1 over K, then

$$n = n_1 + n_2 + ... + n_{k_n(K)}. \tag{21}$$

(Because $x-1|x^n-1$, we may assume that $n_1 = 1$; and for even n, because $x+1|x^n-1$, we may also assume that $n_2 = 1$.) By chapter 9, theorem 5, the isomorphism (4) carries over these degrees. Thus $n_1, n_2, ..., n_{k_n(K)}$ are the degrees of the atomic classes of $L_n(K, V)$.

For arbitrary (allowable) K and V we have the cyclic n-gonal classes defined by the $2^{\tau(n)}$ partial products of the product

$$\prod_{d|n} F_d(x). \tag{22}$$

These '*ever present*' *cyclic n-gonal classes* form a Boolean subalgebra $L_n^\circ(K, V)$ of $L_n(K, V)$. The minimal case $L_n(K, V) = L_n^\circ(K, V)$ appears precisely when (22) is the prime factorization of x^n-1 in $K[x]$, thus when

(i) $k_n(K) = \tau(n)$.

In general

$$\tau(n) \le k_n(K) \le n, \tag{23}$$

and so the maximal case

(ii) $k_n(K) = n$

appears precisely when $x^n - 1$ factors in $K[x]$ into linear factors.

In case (i), (21) becomes the equation

$$n = \sum_{d \mid n} \varphi(d),$$

and in case (ii), $n = 1 + 1 + \ldots + 1$. In chapters 10 and 11 we have discussed what the atomic classes look like in these two extreme cases.

In this chapter a new point of view has emerged through the concept of *anticyclic cyclic n-gonal classes*. By theorem 1 these classes form, for given K and V, a Boolean algebra $L_n^*(K, V)$, and, since they are defined by symmetric and antisymmetric divisors of $x^n - 1$ from $K[x]$ (theorem 2), their number is independent of V. Let the number of atomic classes of $L_n^*(K, V)$ be denoted by $k_n^*(K)$; the cardinality of $L_n^*(K, V)$ is then $2^{k_n^*(K)}$.

The 'ever present' cyclic n-gonal classes are anticyclic: $L_n^\circ(K, V)$ is a Boolean subalgebra of $L_n^*(K, V)$ (see §1, end). Therefore $\tau(n) \le k_n^*(K)$ always holds, and the minimal case $k_n^*(K) = \tau(n)$ appears exactly when $L_n^*(K, V) = L_n^\circ(K, V)$. (A statement equivalent to $k_n^*(K) = \tau(n)$ is: the cyclotomic polynomials $F_d(x)$ with $d \mid n$ have, in $K[x]$, no non-trivial symmetric or antisymmetric divisors.) $k_n^*(K)$ has the greatest possible value precisely when $x^n - 1$ factors, in $K[x]$, into symmetric and antisymmetric factors of degrees 1 and 2, and so for these divisors (21) has the form

$$n = \begin{cases} 1 + 2 + 2 + \ldots + 2 & \text{for odd } n, \\ 1 + 1 + 2 + 2 + \ldots + 2 & \text{for even } n, \end{cases}$$

thus $k_n^*(K) = 1 + [n/2]$; this is the case discussed at the end of §3. In general

$$\tau(n) \le k_n^*(K) \le 1 + [n/2]. \tag{24}$$

$L_n^*(K, V)$ is a Boolean subalgebra of $L_n(K, V)$, and thus $k_n^*(K) \le k_n(K)$ always holds. Of special interest are *the cases where $L_n(K, V) = L_n^*(K, V)$*, i.e., *all cyclic n-gonal classes are anticyclic*; here it is sufficient that the atomic classes of $L_n(K, V)$ are anticyclic. A characterization of these cases is that $k_n(K) = k_n^*(K)$. (An equivalent condition is that all divisors of $x^n - 1$ in $K[x]$ are symmetric or antisymmetric, and it is sufficient that this holds for prime divisors.) Here there are also two extreme cases. The lower one, in which

$$\tau(n) = k_n^*(K) = k_n(K), \quad \text{and so } L_n^\circ(K, V) = L_n^*(K, V) = L_n(K, V)$$

is identical with case (i) which we can consider now as well known. There remains still

(iii) $k_n(K) = k_n^*(K) = 1 + [n/2]$

as a particularly distinctive possibility for the Boolean algebra of cyclic n-gonal classes.

The fields K over which case (i) appears for every n are those over which the cyclotomic polynomials are irreducible. The fields over which case (ii) appears for every n are those which contain the nth roots of unity for every n. The fields over which case (iii) appears for every n are those over which all cyclotomic polynomials except $F_1(x)$ and $F_2(x)$ factor into irreducible quadratic symmetric factors. Maximal ordered fields have this property as we shall now show.

5

REAL COMPONENTS OF AN n-GON

LEMMA *An ordered field contains no roots of unity other than* 1, -1.

PROOF Let K be an ordered field. We wish to show that no polynomial $x^n - 1$ in K has a root different from 1 and -1. First let n be even. Since in the factorization

$$x^n - 1 = (x-1)(x+1)(1 + x^2 + x^4 + \ldots + x^{n-2})$$

the third factor is positive definite, $x^n - 1$ has in K only the roots 1 and -1. If $n = u$ is odd, then, because $x^u - 1 | x^{2u} - 1$, $x^u - 1$ has in K at most the roots 1 and -1. But since $(-1)^u - 1 = -1 - 1 \neq 0$, it has only the root 1.

Now let K be a *maximal ordered field*.[3] The irreducible polynomials of $K[x]$ are the linear polynomials and the definite quadratic polynomials. The cyclotomic polynomials $F_n(x)$ with $n \geq 3$ have, as the lemma above shows, no linear factors and therefore factor into irreducible definite quadratic factors. These factors are symmetric.

For, if $x^2 - cx + d$ is such a factor, then d is an nth root of unity, and thus by the lemma $d = 1$ or $d = -1$. If $d = -1$, then $x^2 - cx + d$ has value -1 at 0, and so is negative definite. Thus in particular it has a negative value at 1 and -1; thus $1 - c - 1 < 0$ and $1 + c - 1 < 0$ which is impossible. Thus $d = 1$ and $x^2 - cx + d$ is symmetric.

The prime factorization of $x^n - 1$ over K can, therefore, with the notation of

3 N. Bourbaki, *Eléments de Mathematique. Algèbre* (Paris, 1952), chap. 6. The maximal ordered fields are, in another terminology (e.g., B.L. van der Waerden, *Modern Algebra* I (New York, 1949), chap. 9), the *real closed fields* with their uniquely determined ordering.

the lemma from §3, be written in the form (19), and (20) is then the decomposition of $\mathfrak{A}_n$ into atomic cyclic classes. Thus we have

THEOREM 9   *Let K be a maximal ordered field. For a given n there are (besides the class of trivial n-gons) $(n-1)/2$ or $n/2$ atomic cyclic n-gonal classes over K, according to whether n is odd or even. These classes consist of affinely regular n-gons with centre of gravity **o**, or affinely regular d-gons with centre of gravity **o**, counted n/d times, where d represents all non-trivial divisors of n. All cyclic n-gonal classes over K are anticyclic.*

In particular, if K is the field R of real numbers, then

$$2 \cos 2\pi/n = e^{i(2\pi/n)} + e^{-i(2\pi/n)}$$

is a number c in the sense of the lemma from §3, since it is a sum of two inverse primitive nth roots of unity; then

$$c_k = 2 \cos k(2\pi/n) \quad (k = 0, 1, 2, ...).$$

For $n = 1, 2, 3, 4, 6$ the prime factorization of $x^n - 1$ and the decomposition of $\mathfrak{A}_n$ into atomic classes are the same over R as over Q. In these cases we have only the cyclic classes which already exist over Q. For $n = 5$ (see figures 92, 93) the only atomic cyclic class over Q, besides the class of trivial 5-gons, is the class of all 5-gons with centre of gravity **o**; over R, however, this class is the direct sum of two atomic affinely regular zero-point classes. These classes are defined by the cyclic systems of equations:

$$a_1 - 2 \cos \tfrac{2}{5}\pi a_2 + a_3 = o, \;...; \quad a_1 - 2 \cos \tfrac{4}{5}\pi a_2 + a_3 = o, \;.... \qquad (25)$$

In particular, over the real affine plane $(V = R^2)$, every 5-gon with centre of gravity **o** is uniquely representable as a sum of 5-gons from the classes given by (25), and thus, we can say, as sum of an affinely regular pentagon with centre of gravity **o**, and an affinely regular pentagram with centre of gravity **o** (figure 94).

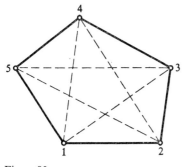

Figure 92

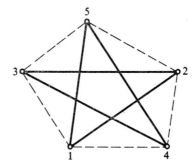

Figure 93

G

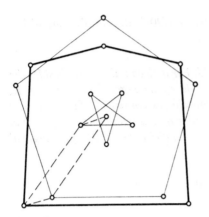

Figure 94

The factorization (19) of $x^n - 1$ which is trivial for $n = 1, 2$ exists for $n \geq 3$ over any field K (with char $K \nmid n$) over which $F_n(x)$ has a quadratic symmetric divisor $x^2 - cx + 1$; but this is not necessarily the prime factorization of $x^n - 1$ over K. $\mathfrak{A}_n$ is then directly representable in the form (20); the terms – the class of trivial n-gons and the symmetric zero-point classes of degrees 1 and 2 – are then the atomic classes of the Boolean algebra of *anticyclic* cyclic n-gonal classes. The components of an n-gon A, which come from this direct-sum representation of $\mathfrak{A}_n$, are called the *real components of A*.

For $n = 1, 2$ the rational, real, and complex components of an n-gon coincide, but for $n = 3, 4, 6$ only the rational and real components coincide.

The real components of an n-gon are affinely regular n-gons or affinely regular d-gons counted several times for proper divisors d of n; except for the trivial centre-of-gravity n-gon, they have centre of gravity o.

Of course, there are more questions connected with the concept of affinely regular n-gons. For example, we should analyse more closely the relation between this concept of an affinely regular n-gon as defined in §3 and the geometric idea of an affine image of a regular n-gon. However every book must come to an end, and so we shall break off at this point.

Exercises

1 Under the assumptions of the lemma in §3 the cyclic projection which maps $\mathfrak{A}_n$ onto the cyclic class defined by the divisor $x^2 - cx + 1$ of $F_n(x)$ is

$$(1/n) \sum_{k=0}^{n-1} c_k \zeta^k.$$

Determine the cyclic projections of $\mathfrak{A}_n$ onto the cyclic classes defined by the divisors $x^2 - c_k x + 1$ of $x^n - 1$ $(k = 1, 2, ..., [(n-1)/2])$.

2 COSINE POLYNOMIAL   Let K be a field whose characteristic does not divide n, and w a primitive nth root of unity from the splitting field of $x^n - 1$ over K. The values

$$w^k + w^{-k} \quad \text{with } k = 0, 1, \ldots, [n/2] \tag{*}$$

are, in the special case $K = R$, twice the cosines of the angles $j(2\pi/n)$ for $j = 0, 1, \ldots, [n/2]$. If we define the monic polynomial whose roots are precisely the values (*) as the nth *cosine polynomial* $C_n(x)$, then we obtain:

$$C_1(x) = x - 2,$$
$$C_2(x) = (x-2)(x+2),$$
$$C_3(x) = (x-2)(x+1),$$
$$C_4(x) = (x-2)(x+2)x,$$
$$C_5(x) = (x-2)(x^2+x-1),$$
$$C_6(x) = (x-2)(x+2)(x+1)(x-1).$$

In general, the nth cosine polynomial $C_n(x)$ has a representation:

$$C_n(x) = \frac{1}{2^{(n-1)/2}} (x-2) \sum_{j=0}^{(n-1)/2} \binom{(n-1)/2}{j} x^{(n-1)/2-j}(x-2)^{[j/2]}(x+2)^{[(j+1)/2]}$$
for odd n,

$$C_n(x) = \frac{1}{2^{(n-2)/2}} (x-2)(x+2) \sum_{j=0}^{[(n-2)/4]} \binom{n/2}{2j+1} x^{(n-2)/2-2j}(x-2)^j(x+2)^j \text{ for even } n.$$

The coefficients of the nth cosine polynomial $C_n(x)$ lie in the prime field of K and are integral rationals when char $K = 0$.

3 KINDER, *Symmetric Cyclic Mappings*

(a) $f(\zeta) \to f(\zeta^{-1})$ is an involutory automorphism of the K-algebra $K[\zeta]$.

(b) The fixed elements of this automorphism (they might be called *symmetric cyclic mappings*) form the subalgebra $K[\eta]$ of $K[\zeta]$ ($\eta := \zeta + \zeta^{-1}$). Its rank is $1 + [n/2]$.

(c) The minimal polynomial of the endomorphism η of V^n is the nth cosine polynomial $C_n(x)$: $K[\eta] \cong K[x]/(C_n(x))$.

(d) The Boolean algebra $E(K[\eta]), \circ, \cdot$ of idempotents of $K[\eta]$ is a Boolean subalgebra of the Boolean algebra $E(K[\zeta]), \circ, \cdot$ of cyclic projections.

(e) The anticyclic cyclic n-gonal classes are the kernels of the symmetric cyclic mappings; for symmetry of Ker $f(\zeta)$ is equivalent with $f(\zeta) \sim f(\zeta^{-1})$ (chapter 6, theorem 5); idempotence of $f(\zeta)$ is equivalent with that of $f(\zeta^{-1})$ (see (a)); and associated idempotents are equal (§6.1, lemma).

(f) The Boolean algebra of anticyclic cyclic n-gonal classes is isomorphic to the lattice of divisors of $C_n(x)$. For $n \neq 1$, there are non-trivial affinely regular n-gons precisely when $C_n(x)$ factors into linear factors.

4 KINDER   In connection with chapter 9, exercise 14, show that, when w is a primitive nth root of unity in an extension field of K:

(a) all cyclic n-gonal classes are anticyclic if and only if w and w^{-1} are conjugate over K (e.g., both these conditions are not fulfilled when $n = 8$ and $K = Q(i)$, the field of Gaussian numbers);

(b) if n is a power of an odd prime, then all cyclic n-gonal classes are anticyclic if and only if $[w:K]$ is even (cf. appendix 2, §4, exercise 2).

5 α-n-GONS Let n be a natural number, V a vector space over a field K satisfying the basic assumptions from §1.1. $\mathfrak{A}_n$ may be interpreted as an End(V)-module. Each endomorphism β of V into itself induces a linear transformation of $\mathfrak{A}_n$ into itself:

$$\beta(a_1, a_2, ..., a_n) = (\beta a_1, \beta a_2, ..., \beta a_n).$$

If $\mathfrak{C}$ is a cyclic class of n-gons, then

$$\text{End } (V)\mathfrak{C} = \mathfrak{C}.$$

If dim $V \geq$ deg $\mathfrak{C}$, then there exists an n-gon $A \in \mathfrak{C}$ with

$$\text{End}(V)A = \mathfrak{C}.$$

Limiting ourselves to automorphisms α of order n of V onto itself, we have $\alpha^n = 1$. For a fixed automorphism α of order n of V onto itself we consider the set of α-n-gons

$$\mathfrak{A}_\alpha = \{(a, \alpha a, ..., \alpha^{n-1}a):a \in V\}.$$

If $m(x) \in K[x]$ is the minimal polynomial of α, then $m(x) \mid x^n - 1$, and

(i) $\mathfrak{A}_\alpha \subseteq \text{Ker } m(\zeta)$,

(ii) $\text{End}(V)\mathfrak{A}_\alpha = \text{Ker } m(\zeta)$.

But, in general, every cyclic class $\mathfrak{C} = \text{Ker } t(\zeta)$ with $t(x)|x^n - 1$ does not have a representation (ii). Nevertheless, if dim $V \geq n$ and $\mathfrak{C}$ is a free cyclic class, then there exists an automorphism α of order n of V onto itself with $\mathfrak{C} = \text{End}(V)\mathfrak{A}_\alpha$. For the representation (ii) can one limit oneself to the idempotent endomorphisms of V into itself?

REFERENCES FOR AFFINELY REGULAR n-GONS

O. Bottema, *De elementaire meetkunde van het platte vlak* (*Groningen*, 1938).

H.S.M. Coxeter, 'Affinely Regular Polygons,' *Abh. Math. Sem. Univ. Hamburg*, vol. 34 (1969).

D. Ruoff and J.R. Shilleto, 'The relationship of affinely regular n-gons to Bachmann and Schmidt's decomposition theory,' forthcoming.

The contents of chapters 10–12 are further developed systematically in H. Kinder, 'Eine geometrische Interpretation der Galoisgruppe von $x^n - 1$,' *Vortrag in Oberwolfach*, 29 May 1969.

Appendices

1
Lattices

by HENNER KINDER

A lattice is a set L with two operations $\sqcup$ and $\sqcap$ (mappings of $L \times L \to L$) and a binary relation $\le$ (subset of $L \times L$) satisfying the following axioms:

Axiom 1 (*Associativity*)

$$(a \sqcup b) \sqcup c = a \sqcup (b \sqcup c),$$
$$(a \sqcap b) \sqcap c = a \sqcap (b \sqcap c).$$

Axiom 2 (*Commutativity*)

$$a \sqcup b = b \sqcup a,$$
$$a \sqcap b = b \sqcap a.$$

Axiom 3 (*Absorption*)

$$a \sqcup (a \sqcap b) = a,$$
$$a \sqcap (a \sqcup b) = a.$$

Axiom 4

$$a \le b \Leftrightarrow a \sqcap b = a.$$

REMARK TO AXIOM 4   If two operations $\sqcup$ and $\sqcap$ on a set L satisfy axioms 1, 2, and 3, then L, $\sqcup$, $\sqcap$, $\le$ is a lattice if $\le$ is defined in conformity with axiom 4.

The following axioms are equivalent to axioms 1 to 4:

Axiom 1' (*Reflexivity*)

$$a \le a.$$

Axiom 2' (*Transitivity*)

$$a \le b \wedge b \le c \Rightarrow a \le c.$$

Axiom 3' (*Antisymmetry*)

$$a \le b \wedge b \le a \Rightarrow a = b.$$

Axiom 4' (*Supremum*)

$$a, b \leq c \Leftrightarrow a \sqcup b \leq c.$$

Axiom 5' (*Infimum*)

$$c \leq a, b \Leftrightarrow c \leq a \sqcap b.$$

EXAMPLES   We shall now give a series of examples of lattices, as well as some processes for constructing further lattices from previous ones. The first example is the most important one, since it is the original type of a lattice.

1. The set $\mathfrak{P}(M)$ of all subsets of a set M with operations $\cup$ and $\cap$ and the relation $\subseteq$.

2. A set $\mathfrak{L}$ of subsets of a set M ($\mathfrak{L} \subseteq \mathfrak{P}(M)$) which is closed under intersection, that is, satisfying the following for all subsets $\mathfrak{M}$ of $\mathfrak{L}$:

$$\bigcap \mathfrak{M} = \bigcap_{T \in \mathfrak{M}} T \in \mathfrak{L}.$$

Then every subset A of M '*generates*' a definite element $\langle A \rangle$ of $\mathfrak{L}$, namely the intersection of all elements of $\mathfrak{L}$ containing A; $\mathfrak{L}$, $\langle \cup \rangle$, $\cap$, $\subseteq$ is a lattice. ($\langle \cup \rangle$ denotes the operation $(T, U) \rightarrow \langle T \cup U \rangle$).

Examples 3 to 6 are special cases of example 2.

3. The set of all subgroups of a group.

4. The set of all normal subgroups of a group. In this lattice $T \sqcup U = \langle T \cup U \rangle = T \cdot U$.

5. The set of all R-submodules of an R-module (R a ring). Here $\langle T \cup U \rangle = T + U$.

6. The set $L(R)$ of all two-sided ideals of a ring R. $\langle T \cup U \rangle = T + U$.

7. The set $L(R) \backslash \{(0)\}$ of all ideals distinct from the zero ideal of an integral domain R with $+$, $\cap$, and $\subseteq$.

8. A principal ideal ring is mapped by $a \rightarrow (a) = Ra$ onto its lattice of ideals $L(R)$ (see example 6). The full pre-image of an ideal Ra is the class Ua of ring elements which are associates of a, where U denotes the group of units of R. Therefore $Ra \rightarrow Ua$ is a one–one mapping of $L(R)$ onto the partition of R into classes of associates. Thus we have defined a lattice isomorphism of $L(R)$, $+$, $\cap$, $\subseteq$ if $\sqcup$, $\sqcap$, and $\leq$ are properly defined in this partition of R into classes: for a, $b \in R$, $Ua \sqcup Ub$ denotes the class of greatest common divisors (g.c.d.) of a and b; $Ua \sqcap Ub$ denotes the class of least common multiples (l.c.m.) of a and b, and $Ua \leq Ub$ means that a is divisible by b.

9. Every *strictly ordered set* or *chain* K (characterized by axioms 1', 2', 3' and *comparability*: for all a, $b \in K$ either $a \leq b$ or $b \leq a$) with $a \sqcup b = \max(a, b)$ and $a \sqcap b = \min(a, b)$.

10. The set $E(R)$ of all idempotent elements of a commutative ring R with $a \sqcup b = a \circ b \, (= a + b - a \cdot b)$ and $a \sqcap b = a \cdot b$ (cf. the remark to axiom 4). See also §5.1.

If L, $\sqcup$, $\sqcap$, $\leq$ is a lattice and if we define a relation $\geq$ in L by

$$a \geq b \Leftrightarrow b \leq a,$$

then L, $\sqcap$, $\sqcup$, $\geq$ is likewise a lattice; it is called the *dual lattice* of L, $\sqcup$, $\sqcap$, $\leq$. Thus in lattice theory we have a *principle of duality*:

If a lattice-theoretic theorem S holds in all lattices, then the 'dual' theorem S^*, formed from S by dualization (replacing $\sqcup$, $\sqcap$, $\leq$ by $\sqcap$, $\sqcup$, $\geq$), also holds in all lattices.

PROOF S^* holds in an arbitrary lattice L, $\sqcup$, $\sqcap$, $\leq$ because S holds in L, $\sqcap$, $\sqcup$, $\geq$.

EXAMPLE

11. The lattice dual to the lattice of example 8 consists of classes of associates with the operations l.c.m., g.c.d., and the relation | ('divides').

If $\mathfrak{L}$ is a *family of lattices*, i.e., a mapping $i \to \mathfrak{L}_i = (L_i, \sqcup_i, \sqcap_i, \leq_i)$ of an (index-)set I into a set of lattices, then its product $\prod \mathfrak{L}$ or $\prod_{i \in I} \mathfrak{L}_i$ is a lattice; $\prod \mathfrak{L}$ consists of the set of all mappings a of I into the union set $\bigcup_{i \in I} L_i$ with $a_i \in L_i$ for all $i \in I$ and the operations $\sqcup$, $\sqcap$ and the relation $\leq$, which are defined in the following manner:

$$(a \sqcup b)_i = a_i \sqcup b_i,$$
$$(a \sqcap b)_i = a_i \sqcap b_i,$$
$$a \leq b \Leftrightarrow \forall i \in I: a_i \leq_i b_i.$$

(If $I = \emptyset$, then $\mathfrak{L} = \emptyset$, and thus, obviously, $\bigcup_{i \in L} L_i = \emptyset$, so that $\prod \mathfrak{L}$ consists only of the empty map of $\emptyset$ into itself, and $\emptyset \sqcup \emptyset = \emptyset \sqcap \emptyset$, and $\emptyset \leq \emptyset$.)

EXAMPLE

12. If $\mathfrak{R}$ is a family of commutative rings with index set $I = \{1, 2, ..., n\}$, then by example 10 the mapping $i \to E(\mathfrak{R}_i)$, $\circ$, $\cdot$ is a *family of lattices*; its product is the lattice $E(\mathfrak{R}_1 \oplus ... \oplus \mathfrak{R}_n)$, $\circ$, $\cdot$ of idempotent elements of the direct sum $\mathfrak{R}_1 \oplus ... \oplus \mathfrak{R}_n$.

The following table shows when a lattice T, $\sqcup'$, $\sqcap'$, $\leq'$ is called a $\sqcup$-sublattice, a $\sqcap$-sublattice, or a sublattice of a lattice L, $\sqcup$, $\sqcap$, $\leq$:

T, $\sqcup'$, $\sqcap'$, $\leq'$ is called a ...	... of L, $\sqcup$, $\sqcap$, $\leq$ if $T \subseteq L$ and for all a, $b \in T$ ...
$\sqcup$-sublattice	$a \sqcup' b = a \sqcup b$
$\sqcap$-sublattice	$a \sqcap' b = a \sqcap b$
sublattice	$a \sqcup' b = a \sqcup b \wedge a \sqcap' b = a \sqcap b.$

EXAMPLES

13. In the lattice $\mathfrak{P}(\{1, 2, 3, 4\})$, $\cup$, $\cap$, $\subseteq$ (see example 1), $\{\emptyset, \{1, 2\}, \{1, 3\}, \{1, 2, 3\}\}$ is a $\sqcup$-sublattice.

14. In the lattice of classes of associates of the principal ideal ring Z with l.c.m., g.c.d., (see examples 11 and 8), with $U = \{1, -1\}$, $\{U, U2, U3, U4, U12\}$ is a $\sqcap$-sublattice.

15. The lattice $\mathfrak{L}$, $\langle\cup\rangle$, $\cap$, $\subseteq$ of example 2 is a $\sqcap$-sublattice of the subset lattice $\mathfrak{P}(M)$, $\cup$, $\cap$, $\subseteq$ (example 1).

16. The lattice of R-submodules of an R-module M (example 5) is a sublattice of the lattice of subgroups of the additive group of M (examples 3 and 4).

17. Every subset of a lattice L, $\sqcup$, $\sqcap$, $\leq$ which is closed with respect to the operations $\sqcup$ and $\sqcap$ forms, with the restrictions of $\sqcup$, $\sqcap$, and $\leq$, a sublattice of L, $\sqcup$, $\sqcap$, $\leq$. We obtain special sublattices of L, $\sqcup$, $\sqcap$, $\leq$ by choosing elements a, b from L in the form:

$(a)_{\leq} := \{x : a \leq x\}$ *(upper (principal) ideal)*

$(b)_{\geq} := \{x : b \geq x\}$ *(lower (principal) ideal)*

$[a, b] := \{x : a \leq x \leq b\}$ *(interval)*.

The following example shows a special case of this.

18. In the lattice of classes of associated elements of a principal ideal ring R with l.c.m., g.c.d., and $|$ from example 11 (and example 8), the classes of associates, which consist of divisors of a given ring element m, form the interval $[U, Um]$ which was denoted by $L(m)$ in §7.4.

Let now L, $\sqcup$, $\sqcap$, $\leq$ be a lattice. Let $a < b$ mean that, for elements a, b of L, $a \leq b$ and $a \neq b$ hold; i.e.,

$a < b :\Leftrightarrow a \leq b \wedge a \neq b$.

The symbol $>$ is explained dually.

An element a of L is called a(n) ...	... of L, $\sqcup$, $\sqcap$, $\leq$, if ...
null element	$a \leq x$ for all $x \in L$
universal element	$x \leq a$ for all $x \in L$
atom	$x < a$ for exactly one $x \in L$
anti-atom	$a < x$ for exactly one $x \in L$

By axiom 3′, L, $\sqcup$, $\sqcap$, $\leq$ has at most one null element. If a null element exists, it is usually denoted by 0. The analogous holds for a unit element 1.

If a is an atom, then any element n with $n < a$ is a null element of L, $\sqcup$, $\sqcap$, $\leq$, because for all $x \in L$ we have $x \sqcap a \leq a$, i.e., $x \sqcap a = a$ or $x \sqcap a = n$; in either

case $n \leq x$. This statement may be dualized. Let us reserve the letters p, q, ...
for atoms. The set of all atoms of L, $\sqcup$, $\sqcap$, $\leq$ will be denoted by $\mathfrak{A}(L, \sqcup, \sqcap, \leq)$.

EXAMPLES

19. In $\mathfrak{P}(M)$, $\cup$, $\cap$, $\subseteq$ (see example 1) $\emptyset$ is the null element and M the universal element. The atoms are the subsets of M containing only one element ('singletons'), and their complementary subsets are the anti-atoms.

20. In lattices which occur in algebra atoms are often referred to as minimal elements, and anti-atoms as maximal elements (e.g., subgroups, ideals, etc.).

21. The closed interval [0, 1] with max, min, $\leq$ formed in the field of real numbers (cf. example 9) has the number 0 as null element, the number 1 as universal element, and no atoms or anti-atoms.

22. In the lattice $\{Ua: a \in R\}$, l.c.m., g.c.d., | of example 11, $U = U1$ is the null element, $\{0\} = U0$ the universal element. The atoms are the classes of associates consisting of prime elements, and there are no anti-atoms (if R is not a field).

We obtain a mapping of the lattice L, $\sqcup$, $\sqcap$, $\leq$ into the lattice $\mathfrak{P}(\mathfrak{A}(L, \sqcup, \sqcap, \leq))$, $\cup$, $\cap$, $\subseteq$ of all sets of atoms of L, $\sqcup$, $\sqcap$, $\leq$ if we associate with each $a \in L$ the set of all atoms $p \leq a$, which we shall denote by $[a]$:

$[a]: = \{p: p \in \mathfrak{A}(L, \sqcup, \sqcap, \leq) \wedge p \leq a\}$.

The mapping

$$a \rightarrow [a] \tag{*}$$

can be quite degenerate; for example, if L, $\sqcup$, $\sqcap$, $\leq$ has no atoms, then $[a] = \emptyset$ for all $a \in L$. In example 1 (cf. also example 19) however, (*) is a lattice isomorphism of $\mathfrak{P}(M)$, $\cup$, $\cap$, $\subseteq$ onto $\mathfrak{P}(\mathfrak{A}(\mathfrak{P}(M), \cup, \cap, \subseteq))$, $\cup$, $\cap$, $\subseteq$; for $x \rightarrow \{x\}$ is a bijection mapping M onto $\mathfrak{A}(\mathfrak{P}(M), \cup, \cap, \subseteq)$. Thus we obtain

THEOREM 1   *If, for a lattice L, $\sqcup$, $\sqcap$, $\leq$, there exist any set M and an isomorphism of L, $\sqcup$, $\sqcap$, $\leq$ onto $\mathfrak{P}(M)$, $\cup$, $\cap$, $\subseteq$, then $\mathfrak{A}(L, \sqcup, \sqcap, \leq)$ is such a set and (*) is such an isomorphism. L, $\sqcup$, $\sqcap$, $\leq$ is isomorphic to the lattice of all subsets of a set if and only if the following hold:*

for all $a, b \in L$ $[a] \subseteq [b] \Rightarrow a \leq b$; $\tag{1}$

for every set A of atoms of L, $\sqcup$, $\sqcap$, $\leq$ there exists an $a \in L$ with $A = [a]$. $\tag{2}$

In the following we shall show that a lattice has these properties if and only if it is a complete, atomic, distributive, and complemented lattice, i.e., a complete, atomic Boolean algebra. But first we must clarify these concepts.

An element s of a lattice L, $\sqcup$, $\sqcap$, $\leq$ is called an *upper limit* of a subset A of L if $a \leq s$ for all $a \in A$. In symbols

$A \leq s.$

$g \in L$ is called an *upper bound* or *supremum* of $A \subseteq L$ if

$$A \leq s \Leftrightarrow g \leq s$$

.e., if g is the null element in the lattice of all upper limits of A (by example 17 this is a sublattice of L, $\sqcup$, $\sqcap$, $\leq$). If $A \subseteq L$ contains a supremum, then it contains only one, and we shall denote it by

$$\sqcup A.$$

Using this notation, we can also express the lattice property (1) in the following way: for all $a \in L$, we have

$$a = \sqcup [a].$$

The concepts of *lower limit*

$$s \leq A$$

and *lower bound* (or *infimum*)

$$\sqcap A$$

of A may be defined dually to the concepts of upper limit and upper bound (or supremum) $\sqcup A$ of $A \leq L$, respectively.

By axiom 4' every subset of L with two elements, $A = \{a, b\}$, has a supremum, namely $a \sqcup b$:

$$a \sqcup b = \sqcup \{a, b\}.$$

A lattice is said to be *complete* if each of its subsets contains a supremum. *Every subset A of a complete lattice also has an infimum*

$$\sqcap A = \sqcup \{x: x \leq A\};$$

for since $a \in A \Rightarrow \{x: x \leq A\} \leq a \Rightarrow \sqcup \{x: x \leq A\} \leq a$, $\sqcup \{x: x \leq A\}$ is a lower bound of A, which is obviously greater than all other lower bounds of A. *Thus the dual lattice of a complete lattice is also complete.*

Every complete lattice L, $\sqcup$, $\sqcap$, $\leq$ possesses a null element, namely $\sqcup \emptyset (= \sqcap L)$, and a universal element, namely $\sqcup L (= \sqcap \emptyset)$.

EXAMPLES

23. The lattice $\mathfrak{P}(M)$, $\cup$, $\cap$, $\subseteq$ of example 1 (and so every lattice with properties (1) and (2)) is complete. The supremum of a set of subsets of M is their union, and the infimum is their intersection.

24. The lattices of examples 2 to 6, 8, 11, and 21 are complete.

25. The product of a family of complete lattices is complete.

26. *Every non-empty finite lattice is complete.*

27. An example of an incomplete lattice is the chain Q, max, min, $\leq$ of the rational numbers (cf. example 9).

By a *complemented* lattice we understand a lattice with 0 and 1 in which each element x possesses at least one *complement*, that is a lattice element y with

$$x \sqcap y = 0 \quad \text{and} \quad x \sqcup y = 1.$$

This property of being complemented is obviously a self-dual lattice property, that is, the dual lattice of a complemented lattice is also complemented. To prove that the product of a family of complemented lattices is complemented requires the axiom of choice.

EXAMPLES

28. The lattice $\mathfrak{P}(M)$, $\cup$, $\cap$, $\subseteq$ (and so every lattice with properties (1) and (2)) is complemented, and in fact every subset A of M has exactly one complement, namely the complement set formed in M:

$$M \backslash A = \{x : x \in M \wedge x \notin A\}.$$

29. The lattice of K-submodules of a K-module (K a field: cf. example 5), thus the subspace lattice of a vector space, is complemented, since every basis of a subspace can be extended to a basis of the whole space.

30. In the lattice $\{Ua : a \in R\}$, l.c.m., g.c.d., $|$ of example 11 (and example 22), an interval $[Ua, Ub]$ with $a \mid b$ and $a \neq 0$ is complemented if and only if the ring element b/a is square-free. $Ut \in [Ua, Ub]$ then has the class $U(a \cdot b/t)$ as its (only) complement. In particular, a sublattice $L(m) = [U, Um]$ (cf. example 18) is complemented if and only if m is square-free. (Units count as square-free also.)

31. In the lattice $E(R)$, $\circ$, $\cdot$ of idempotents of a commutative ring R (see example 10), every non-empty interval $[a, b]$ is complemented; the uniquely determined complement of $x \in [a, b]$ is $a - x + b$. If R has a universal element 1, then the 0 and 1 of R are simultaneously the 0 and 1 of $E(R)$, $\circ$, $\cdot$ and $E(R) = [0, 1]$ is complemented.

(0 and 1 are generally the only idempotent elements of R, if every zero divisor of R is nilpotent, as, for example, in the residue class ring $Z/(p^n)$ or in the domain of integers; for from $a^2 = a$ or $a(a-1) = 0$ we have $a = 1$, or a is a divisor of zero and so nilpotent, and, since idempotent, equal to zero.)

Let us now state a few assumptions under which a lattice fulfils condition (2).

THEOREM 2   *If every element x of a complete lattice has exactly one complement x', then for every set A of atoms of the lattice,*

$$A = [\sqcup A].$$

(2')

PROOF $A \subseteq [\sqcup A]$ is trivial. To prove the converse, let $p \in [\sqcup A]$. Since $p \leq \sqcup A$

and $p \sqcap p' = 0$, we do not have that $\sqcup A \le p'$, and so $A \le p'$ does not hold. Thus there exists a $q \in A$ with $q \not\le p'$. We shall show that $p = q \ (\in A)$.

On the one hand, $p' \sqcap q \ne q$, and thus $p' \sqcap q = 0$. On the other hand, $p' \sqcup q \ne p'$, and thus $p' \sqcup q$ is not the complement of p. But

$$p \sqcup (p' \sqcup q) = 1$$

and so

$$p \sqcap (p' \sqcup q) \ne 0,$$

or else $p, p' \sqcup q$ would be complements.
But

$$p \sqcap (p' \sqcup q) = p$$

and hence

$$p' \sqcup q = p \sqcup (p' \sqcup q) = 1.$$

Thus p and q are both complements of p', and so $p = q$.

A lattice is said to be *distributive* when the *distributive law* holds:

$$a \sqcup (b \sqcap c) = (a \sqcup b) \sqcap (a \sqcup c). \tag{$D_{\sqcup}$}$$

In a distributive lattice, the dual law of $(D_{\sqcup})$ *also holds*:

$$a \sqcap (b \sqcup c) = (a \sqcap b) \sqcup (a \sqcap c), \tag{$D_{\sqcap}$}$$

i.e., *the dual lattice* $L, \sqcap, \sqcup, \ge$ *of* $L, \sqcup, \sqcap, \le$ *is also distributive*.

PROOF $a \sqcap (b \sqcup c) = a \sqcap (a \sqcup c) \sqcap (b \sqcup c)$
$\qquad\qquad\quad = ((a \sqcap b) \sqcup a) \sqcap ((a \sqcap b) \sqcup c)$
$\qquad\qquad\quad = (a \sqcap b) \sqcup (a \sqcap c).$

Dualizing the above proof that $(D_{\sqcup}) \Rightarrow (D_{\sqcap})$, we obtain the converse $(D_{\sqcap}) \Rightarrow (D_{\sqcup})$, which is likewise true by the lattice-theoretic principle of duality.

The distributive law $(D_{\sqcup})$ is valid not only in the dual lattice, but also in sublattices of a distributive lattice and products of distributive lattices.

EXAMPLES

32. The lattice $\mathfrak{B}(M)$, $\cup$, $\cap$, $\subseteq$ (and so also every lattice with properties (1) and (2)) is distributive.

33. *The lattice of ideals of a principal ideal ring*, $L(R)$, $+$, $\cap$, $\subseteq$ *is distributive*. To prove $(D_{\sqcap})$, we must prove that

$$Ra \cap (Rb + Rc) \subseteq (Ra \cap Rb) + (Ra \cap Rc).$$

(The other inclusion holds trivially in any lattice). Let $Rb + Rc = Rd$, so that $d = bu + cv$, $b = db'$, $c = dc'$. Then every element from $Ra \cap (Rb + Rc)$ may be written as xa as well as yd, and hence as $ybu + ycv$ with $ybu = ydb'u = xab'u$ which belongs to $Ra \cap Rb$, and $ycv = ydc'v = xac'v$ which belongs to $Ra \cap Rc$.

34. The lattices of example 8 isomorphic to the lattices of example 33, and their dual lattices $\{Ua: a \in R\}$, l.c.m., g.c.d., $|$ of example 11 are also distributive.

35. Every chain (see example 9) is distributive.

36. For idempotent elements a, b, c of a commutative ring R we have $a(b \circ c) = ab + ac - abc = ab + ac - abac = (ab) \circ (ac)$; i.e., the lattice $E(R)$, $\circ$, $\cdot$ of example 10 is distributive (cf. §5.1).

The elements of a distributive lattice with 0 and 1 have at most one complement. This is proved by two applications of the following theorem.

THEOREM 3   *Let* x_1, y_1 *and* x_2, y_2 *be two pairs of complements in a distributive lattice with 0 and 1. Then*

$$x_1 \leq x_2 \Rightarrow y_2 \leq y_1$$

PROOF $y_2 \leq (y_1 \sqcup x_1) \sqcap (y_1 \sqcup y_2) = y_1 \sqcup (x_1 \sqcap y_2) \leq y_1 \sqcup (x_2 \sqcap y_2) = y_1$.

Complemented distributive lattices are called *Boolean algebras*[1] (or *Boolean lattices*). The complement (uniquely determined by theorem 3) of an element x of a Boolean algebra will be denoted in general by x'. *The mapping*

$$x \rightarrow x'$$

is, again by theorem 3, *an isomorphism of the Boolean algebra* L, $\sqcup$, $\sqcap$, $\leq$, 0, $1'$ *onto its dual Boolean algebra* L, $\sqcap$, $\sqcup$, $\geq$, 1, $0'$. Every Boolean algebra is thus self-dual. In particular, the *de Morgan laws* hold:

$$(a \sqcup b)' = a' \sqcap b' \quad \text{and} \quad (a \sqcap b)' = a' \sqcup b'.$$

By a *Boolean subalgebra* of a Boolean algebra L, $\sqcup$, $\sqcap$, $\leq$, 0, $1'$ we mean a sublattice of L, $\sqcup$, $\sqcap$, $\leq$ which contains 0, 1 and x' for each x in it, that is, a subset of L which is closed under $\sqcup$, $\sqcap$, 0, 1 and $'$ together with the restrictions of $\sqcup$, $\sqcap$, and $\leq$ to the subset. Every Boolean subalgebra of a Boolean algebra is evidently a Boolean algebra.

EXAMPLES

37. $\mathfrak{P}(M)$, $\cup$, $\cap$, $\subseteq$ (and so also every lattice with (1) and (2)) is a Boolean algebra (see examples 28 and 32).

38. *The intervals* $[Ua, Ub]$ *described in example 30, and in particular the lattices of divisors* $L(m)$, *represent, in the square-free case, Boolean algebras* (see example 34).

39. Boolean algebras consisting of idempotent ring elements are obtained from examples 31 and 36.

1   George Boole, 1815–64.

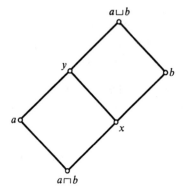

Figure 95

40. In a distributive lattice with 0 and 1 the elements which have a complement form a sublattice (example 17) which is a Boolean algebra.

41. As a further Boolean algebra we should mention the set of all open subsets M of a topological space, which are the interiors of their closures $\bar{M}$, using the interior of $\overline{M \cup N}$ as $M \sqcup N$ and $\cap$, $\subseteq$ as $\sqcap$, $\leq$ respectively.

The *modular law* follows from the distributive law $(D_\sqcup)$

$$a \leq c \Rightarrow a \sqcup (b \sqcap c) = (a \sqcup b) \sqcap c, \qquad (M)$$

which can be seen by applying $(D_\sqcup)$ to the left-hand side of the equation in (M). A lattice satisfying this law is called *modular*. Obviously the dual lattice of a modular lattice is modular. All sublattices and products of modular lattices are modular.

In modular lattices the following 'isomorphism theorem' holds trivially:

For all a, b the $\sqcup$-preserving mapping

$$x \to a \sqcup x \quad (\text{'to join with } a\text{'})$$

of the interval $[a \sqcap b, b]$ into the interval $[a, a \sqcup b]$ and the $\sqcap$-preserving mapping

$$y \to y \sqcap b \quad (\text{'to intersect with } b\text{'})$$

of $[a, a \sqcup b]$ into $[a \sqcap b, b]$ are lattice isomorphisms which are the inverses of each other. That is,

$$a \sqcap b \leq x \leq b \Rightarrow (x \sqcup a) \sqcap b = x \qquad (3)$$

and dually

$$a \leq y \leq a \sqcup b \Rightarrow a \sqcup (b \sqcap y) = y. \qquad (4)$$

Conversely either (3) or (4) implies (M). To prove this, apply, for example, (4) to $y = (a \sqcup b) \sqcap c$.

EXAMPLES

42. The lattice of normal subgroups of a group G (example 4) is modular. In fact, more generally,

$$A \subseteq C \Rightarrow A \cdot (B \cap C) = (A \cdot B) \cap C$$

holds for any three subsets A, B, C of G with $A^{-1} \subseteq A$ and $C \cdot C \subseteq C$. (The proof is trivial, using $A^{-1}C \subseteq AC \subseteq CC \subseteq C$.) In particular, the subgroup lattice of an Abelian group is modular and so (see example 16) also the lattice of R-submodules of an R-module. Thus the expression 'modular.'

43. The $\sqcap$-sublattice of 5 elements named in example 14 is not modular.

44. The subgroup lattice of the group of units of the residue class ring $Z/(8)$ consisting of 5 elements is modular (see example 42), but not distributive.

We shall now exhibit sufficient conditions for a lattice to have property (1). A lattice in which

$$[a] \subseteq [b] \Rightarrow a \leq b \tag{1}$$

holds at least for the case $[a] = \emptyset$, that is, a lattice in which an element a with $[a] = \emptyset$ must be a null element, is called *atomic*. Obviously every non-empty atomic lattice has a null element.

THEOREM 4   *Every atomic, modular, complemented lattice has property* (1).

PROOF   Let $[a] \subseteq [b]$ and c be a complement of $a \sqcap b$. Then

$$[c \sqcap a] = [c \sqcap a] \cap [b] = [c \sqcap a \sqcap b] = [0] = \emptyset,$$

and thus, by the atomic property, $c \sqcap a = 0$. Then by the modular law we can conclude

$$b \geq a \sqcap b = (a \sqcap b) \sqcup (c \sqcap a) = ((a \sqcap b) \sqcup c) \sqcap a = 1 \sqcap a = a.$$

EXAMPLES

45. $\mathfrak{P}(M)$, $\cup$, $\cap$, $\subseteq$ is atomic, as is every lattice with property (1) (see example 19).

46. The lattice $\{Ua: a \in R\}$, l.c.m., g.c.d., $\mid$ of classes of associates of a principal ideal ring R with group U of units is atomic (see example 22).

47. *Every finite lattice is atomic.*

48. The subspace lattice of a vector space (see example 29) is atomic: its atoms are the one-dimensional subspaces.

49. The lattice of ideals $L(R)$, $+$, $\cap$, $\subseteq$ of a ring R with 1 is, in general, not atomic but always '*anti-atomic*' (i.e., the dual to 'atomic').

In conclusion let us bring together theorems 1 to 4, forming the theorem mentioned earlier:

THEOREM 5 *A lattice is isomorphic to the lattice of all subsets of a set if and only if it is a complete, atomic Boolean algebra.*

COROLLARY *A lattice is isomorphic to the lattice of all subsets of a finite set if and only if it is a finite Boolean algebra.*

REFERENCES

G. Birkhoff, *Lattice Theory* (New York, 1961).
H. Hermes, *Einführung in die Verbandstheorie* (Berlin, etc., 1967).
S. MacLane and G. Birkhoff, *Algebra* (New York, 1967).
D. E. Rutherford, *Introduction to Lattice Theory* (Edinburgh and London, 1965).

2
Cyclotomic polynomials

by ECKART SCHMIDT

1

ROOTS OF UNITY

Let n be a natural number and K a field whose characteristic does not divide n.

An element w of an extension field of K is called an *nth root of unity* if w is a root of the polynomial $x^n - 1$.

Because of the assumption about the characteristic of K, the polynomial $x^n - 1$ is relatively prime to its derivative $n \cdot x^{n-1}$. Therefore $x^n - 1$ is square-free and has no multiple roots in any extension field of K (see chapter 6, theorem 3).

Thus a splitting field of $x^n - 1$ over K contains exactly n nth roots of unity. They form a group with respect to multiplication. This group is cyclic, since it is a finite subgroup of the multiplicative group of a field.[1] Its generating elements are called *primitive nth roots of unity*. If w is a primitive nth root of unity, then

$$w, w^2, \ldots, w^{n-1}, w^n = 1$$

are all the nth roots of unity. w^l is a primitive root of unity if and only if $(l, n) = 1$. The number of primitive nth roots of unity is equal to $\varphi(n)$, the value of the Euler function which gives the number of natural numbers l relatively prime to n, with $l \leq n$.

In the cyclic group of nth roots of unity every element has as order a divisor d of n, and so is a primitive dth root of unity. Thus the set of nth roots of unity consists precisely of the primitive dth roots of unity where d runs through all divisors of n.

The splitting fields of $x^n - 1$ over the prime fields are called *nth cyclotomic fields* of characteristic zero or p, according to the characteristic of the prime field (p is a prime with $p \nmid n$). They arise from the prime fields by adjoining a primitive nth root of unity.

The field of complex numbers contains all nth roots of unity for every n.

$$w = e^{2\pi i/n}$$

is a primitive nth root of unity and the nth roots of unity $w, w^2, \ldots, w^{n-1}, w^n = 1$

---

[1] E. Artin, *Galois Theory* 2nd ed., theorem 20 (Notre Dame, 1944).

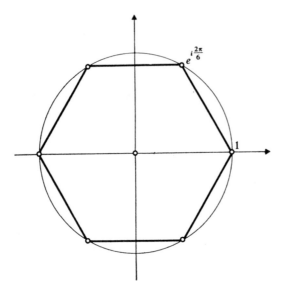

Figure 96

are the vertices of a regular n-gon inscribed in a unit circle in the plane of Gaussian numbers (see figure 96).

Exercises

1 The sum of the nth roots of unity is equal to 0 for $n \neq 1$; the sum of the primitive nth roots of unity is equal to $-\mu(n)$, where μ is the Möbius function. The product of the nth roots of unity is $(-1)^{n+1}$; the product of the primitive nth roots of unity is -1 when $n = 2$ and $+1$ when $n \neq 2$.

2 If the characteristic of K is a prime number p and $n = mp$, then $x^n - 1 = (x^m - 1)^p$, and every nth root of unity is an mth root of unity.

2
CYCLOTOMIC POLYNOMIALS

Let us continue to assume that char $K \nmid n$. By the nth *cyclotomic polynomial* $F_n(x)$ we mean the monic polynomial whose roots are exactly the primitive nth roots of unity, i.e., the polynomial

$$F_n(x) = \prod_w (x - w) \quad (w \text{ runs through all the primitive } n\text{th roots of unity}).$$

The nth cyclotomic polynomial has degree $\varphi(n)$ and is first defined over a splitting field of $x^n - 1$.

Since the nth roots of unity are precisely the primitive dth roots of unity, where d runs through all the divisors of n, and since the primitive dth roots of unity are precisely the zeros of the dth cyclotomic polynomials, we have

$$x^n - 1 = \prod_{d|n} F_d(x). \tag{1}$$

Formula (1) determines $F_n(x)$ uniquely. First $F_1(x) = x - 1$. If $F_d(x)$ is known for all $d \parallel n$, then $F_n(x)$ can be found by division according to the algorithm in the integral domain of polynomials in the indeterminant x; that is, the coefficients of $F_n(x)$ are in the prime field of K and, in the case char $K = 0$, are integral.

In particular, the coefficients of the nth cyclotomic polynomials lie in the prime field of K, and it is sufficient to consider cyclotomic polynomials over prime fields.

All cyclotomic polynomials are irreducible over the field of rational numbers,[2] and (1) is the factorization of $x^n - 1$ into irreducible factors.

EXAMPLE

$$F_1(x) = x - 1,$$
$$F_2(x) = x + 1,$$
$$F_3(x) = x^2 + x + 1,$$
$$F_4(x) = x^2 + 1,$$
$$F_5(x) = x^4 + x^3 + x^2 + x + 1,$$
$$F_6(x) = x^2 - x + 1.$$

The homomorphism of $\mathbf{Z}$ onto $\mathbf{Z}/(p)$ (p a prime number not dividing n), extended to the associated polynomial rings, maps the nth cyclotomic polynomial over the field of rational numbers onto the nth cyclotomic polynomial over the prime field of characteristic p.

From formula (1), with the help of the Möbius inversion formula,[3] we obtain

$$F_n(x) = \prod_{d|n} (x^d - 1)^{\mu(n/d)} \tag{2}$$

Here μ is the Möbius function defined for natural numbers m as follows:

$$\mu(m) = \begin{cases} 1 & \text{if } m = 1, \\ (-1)^r & \text{if } m \text{ is the product of } r \text{ distinct prime factors}, \\ 0 & \text{if } m \neq 1 \text{ is not square-free}. \end{cases}$$

EXAMPLE   For a prime number p

$$x^p - 1 = F_p(x)F_1(x)$$

---

2 See, e.g., B.L. van der Waerden, *Modern Algebra* I (New York, 1949), §36.
3 See, e.g., B.H. Hasse, *Vorlesungen über Zahlentheorie* (Berlin, 1950), §4.7.

and thus

$$F_p(x) = (x^p - 1)/(x-1) = x^{p-1} + x^{p-2} + \ldots + x + 1.$$

If n^* is the square-free kernel of n ($1^* = 1$), then

$$F_n(x) = F_{n^*}(x^{n/n^*}). \tag{3}$$

This result is obtained from (2) by bearing in mind that the Möbius function has value 0 for a non-square-free argument. It is then sufficient to determine cyclotomic polynomials for square-free indices.

The cyclotomic polynomials satisfy the *recursion formulae*

$$
\begin{aligned}
F_{np}(x) &= F_n(x^p) && \text{if } p \text{ is a prime and } p \mid n, \\
F_{np}(x) &= (F_n(x^p))/(F_n(x)) && \text{if } p \text{ is a prime and } p \nmid n.
\end{aligned}
\tag{4}
$$

To prove this write (4) in the form

$$F_{np}(x) = F_n(x^p) \quad \text{for } p \mid n$$

and

$$F_{np}(x)F_n(x) = F_n(x^p) \quad \text{for } p \nmid n.$$

The left and right sides are monic polynomials of the same degree and with the same zeros.

Using (4), we can calculate successively the cyclotomic polynomials of arbitrary index. We commence with

$$F_1(x) = x - 1 \quad \text{and} \quad F_p(x) = x^{p-1} + x^{p-2} + \ldots + x + 1.$$

The cyclotomic polynomial $F_1(x) = x-1$ *is antisymmetric, while on the other hand all cyclotomic polynomials* $F_n(x)$ *with* $n \neq 1$ *are symmetric*, i.e.,

$$F_n(x) = x^{\varphi(n)} F_n(x^{-1}) \quad \text{if } n \neq 1. \tag{5}$$

(See §12.1 for the definitions.)

By (3) it is sufficient to prove (5) for square-free n. Obviously $F_p(x)$ (p a prime) is symmetric. If $F_n(x)$ ($n \neq 1$) is symmetric and p a prime which does not divide n, then by (4) $F_{np}(x)$ is the quotient of symmetric polynomials, and hence again symmetric.

For $F_n(1)$, *the coefficient sum of* $F_n(x)$, *we have*

$$
F_n(1) = \begin{cases}
0 & \text{if } n = 1, \\
p & \text{if } n = p^k \quad (p \text{ a prime}, k \neq 0), \\
1 & \text{otherwise.}
\end{cases}
\tag{6}
$$

By (3) we need only consider the proof for square-free n. Obviously the assertion is true for $n = 1$ and $n = p$ (prime number). If the assertion is true for $n \neq 1$, then by (4)

$$F_{nq}(1) = (F_n(1^q))/(F_n(1)) = 1$$

for a prime $q \nmid n$.

If $n \neq 1$ is odd, then

$$F_{2n}(x) = F_n(-x). \tag{7}$$

By (3) and (4) it is sufficient to prove (7) for odd primes p:

$$F_{2p}(x) = (x^{2p} - 1)(x - 1)/(x^p - 1)(x^2 - 1) = (x^p + 1)/(x + 1)$$
$$= x^{p-1} - x^{p-2} + \ldots - x + 1 = F_p(-x).$$

Exercises

1 Results (2) and (4) are special cases of the relation

$$F_n(x) = \prod_{t \mid d} F_{n/d}^{\mu(d/t)}(x^t) \quad \text{for } d \mid n.$$

2 Prove (5) using the representation (2).
3 Calculate the value of the nth cyclotomic polynomial $F_n(x)$ at -1.
4 The nth cyclotomic polynomial is irreducible over the mth cyclotomic field of characteristic 0 if and only if $(n, m) \mid 2$.

3
A THEOREM OF REDEI

The nth cyclotomic polynomial, as an integral polynomial, generates a principal ideal in $Z[x]$. For this principal ideal we have the following:

THEOREM[4] *In $Z[x]$ the principal ideal $(F_n(x))$ for $n \neq 1$ is generated by the polynomials $F_p(x^{n/p})$; p runs through all the prime divisors of n.*

Once more the proof may be restricted to square-free n (see (3)). The theorem is proved by induction on the number of prime factors of n. If n is prime, the assertion is true. Let $n = mp$, with $m \neq 1$ and p a prime which does not divide m. Then setting

$$f(x) = F_m(x^p)F_{mp}^{-1}(x) \quad \text{and} \quad g(x) = F_p(x^m)F_{mp}^{-1}(x),$$

4 See L. Rédei, 'Über das Kreisteilungspolynom,' *Acta Math. Acad. Sci. Hung.* 5 (1954), 27–8; and I.J. Schoenberg, 'Note on the cyclotomic polynomial,' *Mathematika* 11 (1964), 131–6.

we have $f(x) = F_m(x) \in Z[x]$ by (4) and

$$g(x) = (x^{mp}-1)/(x^m-1)F_{mp}(x) = \prod_{t\|m} F_{tp}(x) \in Z[x].$$

The roots of $f(x)$ are the primitive mth roots of unity u; the roots of $g(x)$ are the primitive (tp)th roots of unity v, where t runs through all proper divisors of m. Then the resultant R of $f(x)$ and $g(x)$ has the form

$$R = \prod_{u,v} (v-u) = \prod_{u,v} v(1-v^{-1}u).$$

R is a unit in the ring of algebraic integers; v is obviously a unit; $v^{-1}u$ can be considered as a primitive dth root of unity w, with $d \neq 1$ and not a prime power.

The resultant R of two polynomials $f(x)$ and $g(x)$ always has a representation

$$R = F(x)f(x)+G(x)g(x),$$

where the coefficients of $F(x)$ and $G(x)$ are from the ring generated by the coefficients of $f(x)$ and $g(x)$.[5] In our case $R \in Z$ and, since R is, in addition, a unit in the ring of algebraic integers, $R = +1$. Thus

$$(1) = (f(x), g(x)) \quad \text{in } Z[x],$$

or

$$(F_{mp}(x)) = (F_m(x^p), F_p(x^m)) \quad \text{in } Z[x]$$

respectively. Then the above theorem follows by induction.

EXAMPLE For $n = pq$ (p, q distinct primes), a representation of $F_n(x)$ corresponding to the Rédei theorem can be given easily. Let $1 = ap-bq$ with $0<a<q$ and $0<b<p$; then

$$F_{pq}(x) = [(x^{ap}-1)/(x-1)]F_{pq}(x) - [(x^{bq}-1)/(x-1)]xF_{pq}(x)$$
$$= [(x^{ap}-1)/(x^p-1)]F_p(x^q) - [(x^{bq}-1)/(x^q-1)]xF_q(x^p).$$

The factors appearing before the cyclotomic polynomials are integral polynomials.

Exercises

1 $F_6(x) = F_3(x^2)-xF_2(x^3)$, $F_{15}(x) = (x^3+1)F_3(x^5)-xF_5(x^3)$.
2 The coefficients of $F_{pq}(x)$ (p, q prime numbers) are $0, \pm1$ only. Determine the number of positive and negative coefficients.
3 $F_{105}(x)$ is the cyclotomic polynomial of least index whose coefficients are not all $0, \pm1$.

5 See, e.g., van der Waerden, *Modern Algebra* I, §2.7.

4

CYCLOTOMIC POLYNOMIALS OVER FINITE PRIME FIELDS

In this section let n be a natural number and p a prime not dividing n. The *order of p modulo n*, i.e., the least natural number l with $p^l \equiv 1 \pmod{n}$, will be denoted by $\mathrm{ord}(p, n)$. By Fermat's Theorem $\mathrm{ord}(p, n) \mid \varphi(n)$.

The nth cyclotomic polynomial is irreducible over a prime field of characteristic zero, but can be factored over a prime field of characteristic p.

EXAMPLE   The $p-1$ non-zero elements of a prime field of characteristic p are the $(p-1)$th roots of unity. Thus, if n is a divisor of $p-1$, the nth cyclotomic polynomial $F_n(x)$ factors into linear factors over the prime field of characteristic p. Thus for $p = 7$

$$F_1(x) = x-1,$$
$$F_2(x) = x-6,$$
$$F_3(x) = (x-2)(x-4),$$
$$F_6(x) = (x-3)(x-5).$$

$n \mid p-1$ means that $\mathrm{ord}(p, n) = 1$. Generally we have

THEOREM   *An nth cyclotomic field of characteristic p has degree $\mathrm{ord}(p, n)$ over its prime field.*[6]

PROOF   Let K be an nth cyclotomic field of characteristic p, and e the degree of K over its prime field. K is a finite field with p^e elements. The multiplicative group of a field with p^l elements (l a natural number) is a cyclic group of order $p^l - 1$ (cf. §1). Among the finite extension fields of its prime field, K is the extension field with the least number of elements whose multiplicative group contains a subgroup of order n; for such a subgroup consists of the n nth roots of unity. A finite cyclic group contains a subgroup of order n precisely when its order is divisible by n. Therefore $p^e - 1$ is the least of the numbers $p^l - 1$ which is divisible by n.

e is thus the least of the natural numbers l with $n \mid p^l - 1$, i.e., with $p^l \equiv 1 \pmod{n}$. Accordingly $e = \mathrm{ord}(p, n)$.

Every primitive nth root of unity of an extension field of the prime field of characteristic p is, according to this theorem, algebraic of degree $\mathrm{ord}(p, n)$ over the prime field. Therefore

COROLLARY   *Over the prime field of characteristic p the nth cyclotomic polynomial $F_n(x)$ is a product of non-associated irreducible polynomials of degree $\mathrm{ord}(p, n)$. The number of factors is $\varphi(n)/\mathrm{ord}(p, n)$.*

6 See, e.g., L. Rédei, *Algebra* I (Leipzig, 1959), §135.

Exercises (on cyclotomic polynomials over finite prime fields of characteristic p)

1 $F_n(x)$ is irreducible if and only if $n = 4, q^k, 2q^k$ ($k \geq 0$; q an odd prime) and p is a primitive number modulo n.

2 If $p = 2$, $F_7(x) = (x^3 + x + 1)(x^3 + x^2 + 1)$;

   if $p = 13$, $F_7(x) = (x^2 + 3x + 1)(x^2 + 5x + 1)(x^2 + 6x + 1)$;

   if $p = 17$, $F_9(x) = (x^2 + 3x + 1)(x^2 + 4x + 1)(x^2 - 7x + 1)$;

   if $p = 3$, $F_8(x) = (x^2 + x - 1)(x^2 - x - 1)$;

   if $p = 11$, $F_{15}(x) = (x^2 - 2x + 4)(x^2 + 4x + 5)(x^2 + 5x + 3)(x^2 + 3x - 2)$.

3 Determine the number of irreducible factors of $x^{12} - 1$ for $p = 5, 7, 11$.

List of symbols and notations

$A, B, \ldots$ n-gons, $O = (o, o, \ldots, o)$ zero n-gon   p. 13

Cyclic n-gonal classes

$\mathfrak{A}_n$	set of all n-gons p. 13
$\mathfrak{A}_{1,n}$	class of trivial n-gons p. 14
$\overset{\circ}{\mathfrak{A}}_n$	zero isobaric class p. 17
$\overset{\circ}{\mathfrak{C}}$	the zero-point class associated with a free cyclic class p. 53
$\mathfrak{A}_{d,\bar{d}}$	class of d-gons counted $\bar{d}$ times ($n = d\bar{d}$) p. 20
$\mathfrak{A}_n^d$	class of n-gons which are d times isobarically split ($d \mid n$) p. 60
$\mathfrak{R}_n$	class of Q-regular n-gons p. 131
$\mathfrak{R}_6$	class of affinely regular hexagons p. 25

Special cyclic mappings

ζ	$(a_1, a_2, \ldots, a_n) \to (a_2, \ldots, a_n, a_1)$ p. 33
σ	The projection σ maps every n-gon $(a_1, a_2, \ldots, a_n)$ into its centre of gravity n-gon, i.e. the n-gon $(a, a, \ldots, a)$ with $a = (1/n)\Sigma a_i$ p. 16
μ_d (for $d \mid n$)	The omitting averaging projection μ_d maps every n-gon $(a_1, a_2, \ldots, a_n)$ into the d-gon with vertices $(d/n)(a_1 + a_{d+1} + \ldots + a_{n-d+1})$, $\ldots$ counted n/d times; these points are the centres of gravity of the omitting sub-n/d-gons of $(a_1, a_2, \ldots, a_n)$ p. 61
κ_d	The consecutive averaging mapping κ_d maps every n-gon $(a_1, a_2, \ldots, a_n)$ into the n-gon with vertices $(1/d)(a_1 + a_2 + \ldots + a_d)$, $\ldots$; these points are the centres of gravity of d-tuples of consecutive vertices of $(a_1, a_2, \ldots, a_n)$ p. 64

$s(\varphi)$	coefficient sum of the cyclic mapping φ p. 35
$K[\zeta]$	algebra of cyclic mappings p. 35
$E(K[\zeta])$	Boolean algebra of cyclic projections p. 77
Ker φ	kernel of φ p. 33
Im φ	φ-image of $\mathfrak{A}_n$ (or of an arbitrary set M) pp. 33, 35
Fix φ	set of fixed elements under φ p. 35

End $(\mathfrak{A})$ ring of endomorphisms of an abelian group $\mathfrak{A}$ pp. 33, 74

annih $\mathfrak{A}$ annihilator of an R-module $\mathfrak{A}$ pp. 95, 116

ker S kernel of an ideal S p. 116

$a \circ b$: $= a+b-ab$ p. 69

$\tau(n)$ number of divisors of n p. 60

$\varphi(n)$ Euler φ-function   p. 179

$\mu(n)$ Möbius μ-function   pp. 139, 181

$d \parallel n$ d is a proper divisor of n: $d \mid n$ and $d \neq n$ p. 114

Special polynomials

$F_n(x)$ nth cyclotomic polynomial   p. 180

$m_d(x)$ $=(d/n)(1+x^d+x^{2d}+ \ \ldots \ +x^{n-d}) = (d/n)(x^n-1)/(x^d-1) \quad (d \mid n)$

           p. 112

Index

addition   13
- displacement of   57
- with respect to the centre of gravity   56
affinely regular   154
- hexagon   25, 28, 44, 61, 79, 91, 114, 131, 136, 148, 153, 154
Alexandroff   xi
algebra of cyclic mappings   33
algebra of cyclic matrices   50
annihilator, annihilate   95
anticyclic cyclic class   149
antisymmetric polynomial   147
Artin   178
ASO-class   23, 30, 42, 60
associated   81
- zero-point class   53
atom   170
atomic components of an n-gon   88
atomic cyclic class   59
atomic cyclic mapping   88
atomic lattice   177
averaging mapping   61, 64

basic classes   18
Birkhoff   xii, 118, 178
Blaschke   5
Bol   5
Boolean algebra   175
- generated   73
- of anticyclic cyclic n-gonal classes   149
- of cyclic n-gonal classes   87, 120
- of cyclic projections   76
- of idempotent elements of a ring   69
Boolean minimal polynomial   73
Boolean subalgebra   175
Boolean sum   70
Bottema   164
Bourbaki   160

centre of gravity   16
characteristic function   110
Chinese construction   82
Chinese isomorphism   101
Chinese residue theorem   99
Choquet   xi, 47

circumscribed   45
coefficient n-tuple of a divisor of $x^n - 1$ 90
coefficient sum of a cyclic mapping   35
complementary lattice   173
complementary projection   37, 62
complementary subspace   55
σ-complementary   55
complete   172
components, atomic   88
- complex   145
- rational   136
- real   162
congruent modulo (m) 82
consecutive averaging   64
cosine polynomial   163
counted   19, 31, 113
Coxeter   164
cyclic class, atomic   59
- defined by a divisor of $x^n - 1$ 108
- free   17, 53, 112
cyclic class of n-gons   15
cyclic mapping   32
- atomic   88
- isobaric   51
cyclic matrix   21
cyclic projection   39
cyclic reflection   88
cyclic system of equations   14
cyclotomic polynomial   85, 134, 151, 154, 180

decomposition of an n-gon   40, 64, 80, 88, 137, 140, 161
degree of a cyclic class   20, 122
degree of freedom   20, 122
de Morgan laws   70, 175
derivative formula   107
dimension   21, 122
distributive lattice   174
divisor of $x^n - 1$ and cyclic n-gonal classes   87, 89, 108, 112, 120, 123, 135, 141, 150, 154, 158, 162

Euclidean algorithm   107

free cyclic class 17, 52, 112
freely generate 73

Galois correspondence 116
Galois group 128, 164
group algebra 35

Hasse 139, 181
Hermes 178
hexagon 6, 24, 30, 41, 54, 61, 64, 78, 88, 91, 114, 136, 140, 145, 148, 154

ideal-transfer 118
idempotent 37, 69
– modulo (m) 82
idempotent-transfer 96
Im-transfer 74
interval 170
involutory 71
isobaric 16
– cyclic mapping 35, 51
isobaric class 17, 58
isobarically split 60

Jacobsen 69

kernel 33, 37, 116
σ-kernel 51
Kinder xii, 127, 163, 164, 167
Kochendoerffer xii, 81
Kowalsky xii
Kurosch 81

Lang xii
lattice 167
– of divisors 97

MacLane 178
main diagram 121
main theorem 87, 120
minimal polynomial 121
– Boolean 73
Möbius function 139, 181
modular 53, 176
module 105

n-gon 13
n-gonal class. See cyclic

octagon 26, 62
omitting, averaging 61
omitting subpolygon 19
opposite vertices, sides 14
orthogonal 69

parallelogram 3, 15, 29, 45, 49, 57, 66, 91, 126, 131, 148, 153, 154

parallelogram point, fourth 15
$2m$-parallelogram 5, 22, 30, 42, 60, 64, 66, 134
partial sums 71
pentagon 145, 161
pentagram 145, 161
periodic classes 19
point 13
polynomial ring 84
principal ideal domain 81
prism 6, 24, 28, 30, 31, 42, 56, 61, 66, 80, 140
– spanned 79
projection 37
– cyclic 39

Q-regular 131
– components 136
quadrangles 3, 22, 40, 48, 57, 91, 144, 148
quasi-projection 37, 86

R-module 95
rank of a cyclic mapping 122
Rédei 183, 185
reflection, cyclic 88
regular 142, 143
relatively prime jumps 143
root of unity 110, 141, 179
Rutherford 178

Schoenberg 183
side vectors 14
sides 14
spectrum 110
spot 110
square 57, 144, 148
square-free 82, 93, 134, 173, 175
sublattice 169
subpolygon 19
symmetric polynomial 149
σ-complementary 55
σ-kernel 51

Thomsen 5
track 89
transfer 74
triangle 24, 148
trivial n-gon 14
typical atomic class 114

van der Waerden xii, 69, 160, 181, 184
vector space of n-gons 14
vertex 14
vertex scheme 19

w-components 145

w-n-gon 141

$x^n - 1$ 84, 179
– divisor of, and cyclic n-gonal classes 87, 89, 108, 112, 120, 123, 135, 141, 150, 154, 158, 162

Yaglom 4

zero isobaric class 17
zero n-gon 13
zero-point class 18, 53, 112
– associated 54

120473

LIBRARY
OF
MOUNT ST. MARY'S
COLLEGE
EMMITSBURG, MARYLAND

LIBRARY
OF
MOUNT ST. MARY
COLLEGE
EMMITSBURG, MARYLAND

DEC 16 1975